This book surveys the physics of small clusters of particles undergoing vibrations, with applications in nuclear physics and the physics of atomic clusters.

The book begins with a survey of the experimental information on collective vibrations in atoms, metal clusters and nuclei. Next, the book goes on to develop theoretical tools to understand these findings. Special emphasis is placed on the Rayleigh-Ritz principle, the use of sum rules, and the quantum mechanics of mean field theory, known as "RPA". The important vibrational modes observed in the different systems are then discussed, including the dipole mode of oscillation (important in both nuclei and metal clusters), surface modes of higher multipolarities, and the compressional mode. In the last two chapters mechanisms for the damping of vibrational modes and the effects of excitation energy on the modes are described.

This book will be of interest to experimentalists and theorists studying finite quantum systems in nuclear physics, atomic physics or physical chemistry.

T0282295

CAMBRIDGE MONOGRAPHS ON
MATHEMATICAL PHYSICS

General Editors: P. V. Landshoff, D. R. Nelson, D. W. Sciama, S. Weinberg

OSCILLATIONS IN FINITE
QUANTUM SYSTEMS

Cambridge Monographs on Mathematical Physics

A. M. Anile *Relativistic Fluids and Magneto-Fluids*
J. Bernstein *Kinetic Theory in the Early Universe*
G. F. Bertsch and R. A. Broglia *Oscillations in Finite Quantum Systems*
N. D. Birrell and P. C. W. Davies *Quantum Fields in Curved Space*[†]
D. M. Brink *Semiclassical Methods in Nucleus–Nucleus Scattering*
J. C. Collins *Renormalization*[†]
P. D. B. Collins *An Introduction to Regge Theory and High Energy Physics*[†]
M. Creutz *Quarks, Gluons and Lattices*[†]
F. de Felice and C. J. S. Clarke *Relativity on Curved Manifolds*
B. DeWitt *Supermanifolds, second edition*[†]
P. G. O. Freund *Introduction to Supersymmetry*[†]
F. G. Friedlander *The Wave Equation on a Curved Space-Time*[†]
J. A. H. Futterman, F. A. Handler and R. A. Matzner *Scattering from Black Holes*
M. Göckeler and T. Schücker *Differential Geometry, Gauge Theories and Gravity*[†]
M. B. Green, J. H. Schwarz and E. Witten *Superstring Theory, volume 1: Introduction*[†]
M. B. Green, J. H. Schwarz and E. Witten *Superstring Theory, volume 2: Loop Amplitudes, Anomalies and Phenomenology*[†]
S. W. Hawking and G. F. R. Ellis *The Large-Scale Structure of Space-Time*[†]
F. Iachello and A. Arima *The Interacting Boson Model*
F. Iachello and P. van Isacker *The Interacting Boson Fermion Model*
C. Itzykson and J.-M. Drouffe *Statistical Field Theory, volume 1: From Brownian Motion to Renormalization and Lattic Gauge Theory*[†]
C. Itzykson and J.-M. Drouffe *Statistical Field Theory, volume 2: Strong Coupling, Monte Carlo Methods, Conformal Field Theory, and Random Systems*[†]
J. I. Kapusta *Finite-Temperature Field Theory*
D. Kramer, H. Stephani, M. A. H. MacCallum and E. Herlt *Exact solutions of Einstein's Field Equations*
N. H. March *Liquid Metals: Concepts and Theory*
L. O'Raifeartaigh *Group Structure of Gauge Theories*[†]
A. Ozorio de Almeida *Hamiltonian Systems: Chaos and Quantization*[†]
R. Penrose and W. Rindler *Spinors and Space-time, volume 1: Two-Spinor Calculus and Relativistic Fields*[†]
R. Penrose and W. Rindler *Spinors and Space-time, volume 2: Spinor and Twistor Methods in Space-Time Geometry*
S. Pokorski *Gauge Field Theories*[†]
V. N. Popov *Functional Integrals and Collective Excitations*[†]
R. Rivers *Path Integral Methods in Quantum Field Theory*[†]
R. G. Roberts *The Structure of the Proton*
W. C. Saslaw *Gravitational Physics of Stellar and Galactic Systems*[†]
J. M. Stewart *Advanced General Relativity*
R. S. Ward and R. O. Wells Jr *Twistor Geometry and Field Theories*[†]
J. Fuchs *Affine Lie Algebras and Quantum Groups*

[†] Issued as a paperback

OSCILLATIONS IN FINITE QUANTUM SYSTEMS

G. F. BERTSCH

National Superconducting Cyclotron Laboratory, Michigan State University

R. A. BROGLIA

Dipartimento di Fisica, Università di Milano and INFN Sez.Milano,
and The Niels Bohr Institute, University of Copenhagen

CAMBRIDGE
UNIVERSITY PRESS

CAMBRIDGE UNIVERSITY PRESS
Cambridge, New York, Melbourne, Madrid, Cape Town, Singapore, São Paulo

Cambridge University Press
The Edinburgh Building, Cambridge CB2 2RU, UK

Published in the United States of America by Cambridge University Press, New York

www.cambridge.org
Information on this title: www.cambridge.org/9780521411486

First published 1994
This digitally printed first paperback version 2005

A catalogue record for this publication is available from the British Library

Library of Congress Cataloguing in Publication data
Bertsch, George F.
Oscillations in finite Quantum systems / G.F. Bertsch, R.A. Broglia.
    p.   cm. – (Cambridge monographs on mathematical physics)
Includes bibliographical references and index.
ISBN 0 521 41148 3
1. Many-body problem.   2. Oscillations.   3. Atoms.   4. Metal
crystals.   5. Nuclear physics.   6. Mathematical physics.
I. Broglia, R.A.   II. Title.   III. Series.
QC174.17.P7B458   1994
530.1′44–dc20   92-40596   CIP

ISBN-13 978-0-521-41148-6 hardback
ISBN-10 0-521-41148-3 hardback

ISBN-13 978-0-521-01996-5 paperback
ISBN-10 0-521-01996-6 paperback

for Angela, Donatella, Gianandrea and Bettina–RAB

Contents

Preface

One of the fascinating questions in Nature is trying to understand how the properties of macroscopic systems emerge from
the quantal behavior of its constituents. For example, how many
atoms does it take to make a solid? Equally interesting is to ask
how the macroscopic behavior of large systems emerges as a limit
point when one observes the properties of finite systems. A powerful technique, both theoretically and experimentally, to study this
question is to examine the response of the system to weak external
perturbations. The study of this subject is the central theme of the
present monograph.

We start by surveying the experimental information on collective
vibrations in atoms, metal clusters and nuclei. It will be apparent
that the vibrational modes and their frequencies can reveal much
about the nature of the forces acting within the system.

Following the overview, we develop the main tools to provide an
understanding of these findings. We place special emphasis on the
Rayleigh–Ritz principle, the use of sum rules, and the quantum
mechanics of mean field theory, known as 'RPA'.

With the various classical and quantum mechanical tools, we
proceed to discuss the important vibrational modes observed in
the different systems. The dipole mode of oscillation is prominent
in both nuclei and metal clusters, and there are remarkable sim-
ilarities between the two systems—on very different energy scales
of course. Surface modes of higher multipolarity are prominent
features of the nuclear response, and quantum mechanics of the
many-fermion system produces a rather subtle response. The compressional mode is also important in nuclei, because it bears most
closely on the properties of macroscopic nuclear matter.

In the last two chapters we discuss the mechanisms responsible for the damping of vibrational modes and the effects of excitation energy on the modes.

The book is aimed mainly toward students and experimental researchers studying finite quantum systems. We have deliberately avoided technicalities, formalism, and many details. Our emphasis is on the physics of the response of these small systems. In keeping with this spirit, we have not compiled an extensive bibliography, but we hope we have cited enough material, via other books and review articles, for the reader to find more extensive literature. Readers who are interested in carrying out numerical RPA calculations themselves may obtain free copies of relevant programs described in App. E.

We would like to thank A. Bulgac, P.F. Bortignon, D. Nocera, D. Tomanek and K. Yabana for discussions, and in addition J. Foxwell, G. Lazzari and Y. Wang for reading parts of the manuscript. We also thank O. McHarris for the figures, and G. Colò for providing figures for selected topics, and for helping to prepare one of the computer codes for distribution.

1

Introduction

The behavior of particles interacting under the laws of quantum mechanics is a fundamental concern of both physics and chemistry. The equations describing a system of two particles can be solved efficiently to any needed accuracy. When more than two particles interact simultaneously, however, the *ab initio* quantum mechanical calculations offer little insight.

Fortunately, the actual behavior of the system may be quite simple, even when the underlying equations are not. Depending on the physical situation, only a few degrees of freedom may be relevant. If these degrees of freedom involve the motion of all the particles in the system, we call the motion collective. A familiar example of collective motion is sound in solids. Sound waves are described by a field that acts on all particles in the system in the same way, depending only on the location of the particles. The field is particularly simple for infinite systems because the translational invariance permits only a very regular motion in the normal modes, each mode having a definite wavelength. In finite systems the existence of a surface is very important in determining the normal modes, and the modes cannot be described with definite wavelengths. Nevertheless, we shall see in the finite systems examined here that the important modes are often simple to characterize.

In the language of quantum mechanics, motion in a system is usually described in terms of transitions between energy levels. Each transition has its own frequency, corresponding to the energy difference between the two levels. Collective motion makes its appearance in the quantum language when particular states have large transition amplitudes. We will see that it is then often

1

possible to describe the transition rather well, even though many of the details of both the initial and final states are not well known.

If we were to make a complete theory of finite systems, we would have to begin with its equilibrium properties. We would first like to know what the stable structure of the system is. How large is it, where are the particles, what is its binding energy? The answers depend not only on the intrinsic forces but upon the conditions under which the system is studied, whether it is cold and in its ground state, or hot and in some statistical ensemble of states.

In this book we pass over these interesting first questions and jump directly to the question, how does the system respond to the external environment? In particular, if the system is subjected to an external field, how do the particles move and how does the system as a whole absorb energy? In classical physics, this is partly answered by finding the normal modes of the system, the small-amplitude vibrations. One of the fascinating properties of the quantum systems is that the classical modes still set the stage for the quantum motion, even though in principle that motion can be much richer.

In this first chapter, we introduce our subject with an overview of the experimental methods used to study the oscillations of finite quantum systems. Later chapters will discuss the theory, starting from classical concepts appropriate to collective motion. We will then build the quantum theory one step at a time.

1.1   Probing the system with photons

Electromagnetic fields provide one of the most important methods for probing many-particle systems ranging from molecules and atomic clusters on the scale of nanometers, down to nuclei on a scale of femtometers. The electromagnetic interaction is completely known, it is strong enough to produce easily observed effects, and yet it is weak enough for the effect of the interaction to be separated from the intrinsic properties of the system under study.

We must first distinguish between static electromagnetic fields and fields that vary in time. With static fields, one studies polarizabilities and static electromagnetic moments, i.e., properties of the equilibrium state. The time-varying fields are generally more interesting, since they can induce internal motion. Physically, the simplest such field is a free electromagnetic wave, described quantum-mechanically with photons. In the language of photons, one can treat all physical processes using scattering theory. The most important concept is the cross section, which is the probability of the system's absorbing a photon multiplied by the area illuminated by the photon beam. From an experimental side, properties are investigated by measuring the cross sections. A schematic sketch of an experiment to measure a photon cross section is shown in Fig. 1.1. A well-collimated beam of photons passes through a monitor and enters a detector such as a spectrometer. The absorbing material to be investigated is placed in the photon beam between these two instruments, and absorbs some fraction of the beam. The fraction is given by the ratio of photons transmitted, N_f, to photons incident, N_0. The latter might be determined by taking the absorber out of the beam. The cross section relates the attenuation to the areal density of particles in the cell, n_A. This is given by $n_A = L\rho x/A$, where L is Avogadro's number, ρ is the density of the absorber, x is its length, A the atomic or molecular weight. Then the cross section σ is obtained from

$$\sigma = \frac{1}{n_A} \log_e(N_0/N_f) \ .$$

This is the attenuation method for measuring cross sections. A simple example of its use is shown in Fig. 1.2. Here is graphed the transmitted intensity of an infrared beam passing through a gas cell containing the molecule HCl. The attenuation factor, I_f/I_0, is shown as a function of the wavenumber of the photon. Notice that the spectrum consists of a number of narrow lines. These are resonances that correspond to definite transitions between quantum states of the molecule. The HCl molecule is a nearly rigid object at these frequencies; only the quantized states of vibration and rotation play a role in the transition frequencies. The widths of the resonances in this spectrum are extrinsic, associated with instrumental resolution and interactions between the molecules. Other extrinsic effects, such as the Doppler shift of the radiation on a moving molecule, can affect the apparent width as well. The intrinsic width of the resonance is an important property of the molecule itself, but is too small to observe under the experimental conditions of Fig. 1.2. In the usual nomenclature, the width Γ is defined as the interval of energy over which the cross section is

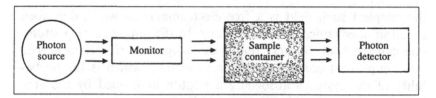

Fig. 1.1. Schematic representation of a photon absorption experiment.

more than half its maximum value. We shall also use the same symbol to express the interval in angular frequency units. The width of a state Γ is related to its mean lifetime τ by the simple formulas

$$\Gamma = \frac{1}{\tau} \text{ (frequency units)}$$

$$\Gamma = \frac{\hbar}{\tau} \text{ (energy units)}.$$

In the unit conversion, Planck's constant $\hbar$ is often conveniently expressed as $\hbar = 6.7 \times 10^{-16}$ eV s. Our next example of photon absorption shows a case where the intrinsic width of the excitation is a significant fraction of the photon's frequency. The system studied is the gadolinium atom, and the photons are in the X-ray region. The absorption cross section for photons in the energy range 120–200 eV is shown in Fig. 1.3. The main feature in the spectrum is a broad, somewhat asymmetric peak in the cross section. The resonance is caused by the single-electron transition from an inner d-shell to the valence f-shell in the atom. The width of the peak is intrinsic to the atom and is associated with the lifetime of an electron to escape from the f-shell. In this atom, there is a small barrier in the f-wave potential that produces a resonant state having a finite width.

In the above examples of a molecule and an atom, the spectra are rather directly related to the single-electron atomic shell physics or to the geometric properties of a quantum vibrator. The next example is from systems of atoms that display more subtle interaction effects between the electrons, namely, metal atom clusters. These clusters have been studied intensively recently (cf. the proceedings of the Konstanz conference (1991) and references therein). In typical experiments the clusters are formed in beams and probed with photons or external perturbations while

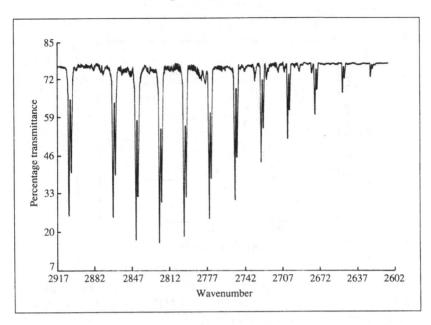

Fig. 1.2. Transmitted intensity of an infrared beam passing through a cell containing gaseous HCl, as a function of photon wavenumber(courtesy of R.A. Bertsch). The lines form a band corresponding to the angular momentum states of the lowest vibrational transition. Wavenumber units n are related to photon wavelength λ and photon energy e_p by $n = 1/\lambda = 8067.5(e_p)_{eV}$ cm^{-1}.

still in the beam. These stringent conditions are needed to study the clusters because they are generally quite fragile and would disappear on contact with a surface or collision with another particle. The photon absorption spectra for the alkali metal cluster Na$_8$ is shown in Fig. 1.4. In this system, there is a single prominent peak in the spectral region associated with the valence electron shells. We shall see in Chap. 5 that the strength of this peak is close to the maximum permitted for a valence transition at this energy; its strength nearly exhausts an energy-weighted sum rule. This implies that the valence electrons from all of the atoms participate together in the resonance. The position of the resonance is different from the single-electron shell energy: in a sodium atom, the valence s to p transition has a wavelength of 589 nm, but in the cluster formed by eight atoms the peak is around 500 nm. The collective interaction effects have shifted the peak downward in wavelength and upward in frequency. If this were a classical sys-

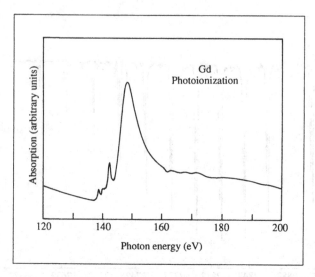

Fig. 1.3. The absorption spectrum of Gd vapor in the range of the transition $4d \rightarrow f$ making use of the 500 MeV synchrotron of Bonn (from Connerade and Pantelouris (1984)).

tem of free electrons, the resonance frequency would be calculable with Newtonian mechanics and electrostatics. The classical theory produces a definite formula, and the corresponding resonance is called the Mie resonance, named after a pioneer investigator of electromagnetic resonances in spherical dielectrics (Mie 1908). In this mode, the electrons simply move back and forth uniformly with a spherical volume, as depicted in Fig. 1.5. In the sodium example, the Mie resonance corresponds to a wavelength of 430 nm, which is smaller than observed. Thus, the finite system behaves somewhere between the limiting extremes of the isolated atom and of a sphere of metallic sodium.

Nuclear physics provides many examples of excitations in which the neutrons and protons making up the nucleus move collectively. For nuclear physics, the convenient length scale is the femtometer, 1 fm $= 10^{-15}$ m, and the corresponding energy scale is MeV, megaelectron volts. Photon absorption experiments still follow the classical set-up of Fig. 1.1. The source of photons is most often bremsstrahlung radiation from very energetic electron beams. To select photons of a given energy range, it is not possible to use diffraction as in the case of optical photons or X-rays. Instead, the individual photons must be 'tagged' by measuring simultaneously

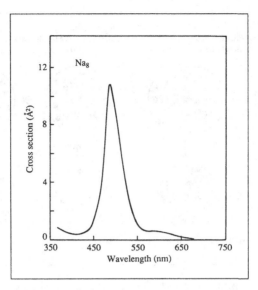

Fig. 1.4. Photoabsorption cross section of Na_8 (in angstroms squared, 1 Å= 10^{-8} cm) as a function of the photon wavelength in nm (Wang et al. (1990)). The clusters were generated by expanding sodium vapor from a high-temperature oven source into a vacuum. The cross section was determined making use of the technique of beam photodepletion. In this method, a pulsed laser beam is directed collinear with but counterpropagating the cluster beam. The beam is detected by ionizing it with another photon source and then accelerating it through a quadrupole mass analyzer. At resonance with the pulsed laser frequency, the cluster beam is depleted due to dissociation of the excited clusters that had absorbed a photon from the laser beam. Note that this experiment measures only the dissociation cross section, which is equal to the absorption cross section only if the cluster has adequate time to decay and nondissociative decay processes are negligible.

the electron that produced the photon and the electron's energy loss. Another difficulty is that the attenuation method for measuring cross sections is less easy to use because only a fraction of the photons are absorbed directly by the nucleus. Most of the beam attenuation comes from interactions with electrons. Nevertheless the attenuation method can be applied, yielding cross sections such as shown in Fig. 1.6. Here we see the photon absorption cross section on the nucleus ^{12}C, measured for photons in the energy range of 15 to 100 MeV. The main feature of the spectrum is a single peak, which is located at an energy near 25 MeV. This is called the giant

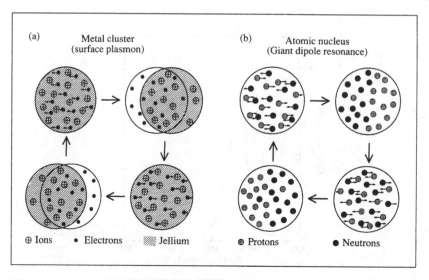

Fig. 1.5. Schematic representation of the giant dipole resonance in atomic nuclei and of the surface plasmon resonance in small metal clusters. The wavelength of the photon exciting these vibrations is large with respect to the diameter of the system. As a result the electric field associated with a passing gamma ray is nearly uniform across the system. In the case of the excitation of the atomic nucleus, the field exerts a force on the positively charged protons, thus separating them from the neutrons. In fact, the neutrons act as having an (effective) negative charge, which oscillates out of phase with respect to the positively charged protons. This is the reason why the giant dipole vibration is an isovector vibration. In the case of metal clusters, the electric field associated with the photon exerts a force on the positively charged ions and an identical force but with opposite direction on the electrons. Because the ions have a mass which is three orders of magnitude larger than that of the electron, the displacement of the electron cloud is much larger than that of the positive background.

dipole resonance. Its strength, as measured by the energy-weighted sum rule, shows that it is a collective excitation involving all of the protons and neutrons. The motion is very similar to that of the Mie resonance. The protons move uniformly back and forth, as shown in Fig. 1.5. The neutrons also move, in an opposite sense from the protons, because, for an internal excitation, the center of mass of the system cannot oscillate, but only recoil. A magnified view of the ^{12}C giant dipole resonance is shown in Fig. 1.7. In this experiment, the cross section for producing neutrons from the

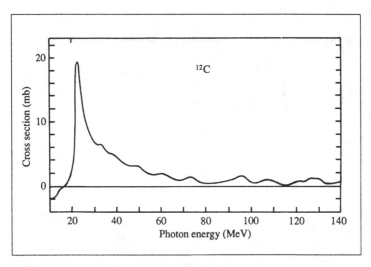

Fig. 1.6. Nuclear photoabsorption cross section of the carbon nucleus, shown as a function of the photon energy. The cross section is given in millibarns (1 bn = 10^{-24} cm^2, 1 mb = 10^{-27} cm^2). The peak at around 23 MeV is due to the giant dipole resonance. The angular frequency of the oscillation is given by $\omega = E/\hbar = (23 \text{ MeV})/(6.7 \times 10^{-22} \text{ MeV s}) = 3 \times 10^{21} \text{ s}^{-1}$, corresponding to an ordinary frequency of $f = \omega/2\pi = 5 \times 10^{20}$ Hz. The data is from Ahrens et al. (1972).

absorbed photons was measured. In contrast, in an attenuation experiment, one measures the total cross section for any beam interaction. The finite width of the resonance is clearly seen, as well as the fact that it is not a smooth function. As we shall see in Chap. 9, in these nuclear systems the width is not due primarily to the escape of particles. It is caused by mixing the collective mode with more complicated internal excitations. The irregularities in these internal states produce the extra structure in this cross section.

It is amusing to look at photon interactions on an even smaller scale, for example, the absorption on a single proton. The experimental spectrum is shown in Fig. 1.8. It shows a resonant peaking similar to that seen for nuclei or larger systems. In this case, the proton behaves as a composite system, and the particles responsible for the resonance are its quark constituents. Although we presented this figure, it is really beyond the domain of our book; the theory is far from adequately developed for quarks and their interactions.

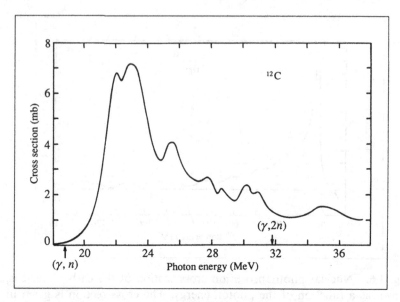

Fig. 1.7. Total photoneutron cross section $(\gamma,\text{total}) = [(\gamma, n) + (\gamma, 2n) + \ldots]$ associated with ^{12}C. The photons in this experiment were produced by an electron beam, and their energy measured ('tagged') by detecting the slowed electrons. The beam passed through a target of ^{12}C, and neutrons produced by the photoabsorption were detected in coincidence with the tagged photons. The giant dipole resonance at 23 MeV is seen in this higher resolution experiment to have a substructure of smaller peaks. The data is from Berman and Fultz (1975).

1.2   A second probe of resonances: inelastic scattering

Resonant photon absorption, which we surveyed in the last section, involves mainly photons whose wavelengths are much larger than the size of the cluster. In the long wavelength limit the electric field is nearly uniform over the entire cluster, and the preferred induced motion is just a uniform displacement of the charges. If we wish to study more complex patterns of motion, the external field must have some spatial variation. Also, fields other than the electromagnetic may be interesting to study in their own right. Such fields can be made by scattering a particle from the system. Indeed, one of the early experiments in quantum mechanics was the famous Franck–Hertz experiment, which produced excitations in atoms by the inelastic scattering of electrons. Of course, that

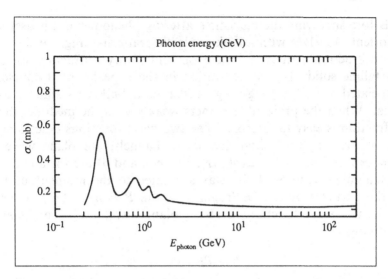

Fig. 1.8. Cross section for photons scattering from protons, shown as a function of the energy of the photon. The strong peak is due to the Δ-resonance, an excited state of the proton (from Hernandez et al. (1990)).

experiment was very crude; the energy transfer to the atom was detected by a decreased current from the slowed electrons.

In modern experiments one measures both the energy loss of the scattered particle and its angular deflection. The energy loss corresponds to the frequency of the excitation in the system as in the photon absorption process, but the angular deflection provides completely new information about the system. The deflection caused by the diffraction of the quantum mechanical waves associated with the particle is related to the spatial distribution of the field acting on the particle. This is a familiar technique for studying bulk materials; the distribution of atoms in a crystal is inferred from X-ray or neutron diffraction patterns.

According to the laws of quantum mechanics, the field produced by a scattering particle has a spatial variation given by the product of the initial and final wave functions of the particle. A mode will be more or less difficult to excite depending on how the motion of the particles in the mode fits together with the spatial variation of the field. Since the field can be varied by looking at scattering in different directions, more detailed information about the excitation mode can be extracted.

Before surveying the inelastic scattering phenomena, it is useful to orient ourselves with a look at elastic scattering. Fig. 1.9 shows the diffraction pattern for neutron and X-ray diffraction on a crystalline solid. The concentration in sharp peaks is, of course, associated with the long-range order of a bulk crystalline material. When the projectile interacts weakly with the medium, the diffraction is easy to interpret. The two wave functions describing the electron in the initial and final channels, are plane waves $\psi_i = \exp(ip_ir)$ and $\psi_f = \exp(ip_fr)$, where p_i and p_f are the angular wavenumber vectors of the waves, related to the momentum P of the particles by the DeBroglie relation, $P = \hbar p$. The product of the two wave functions is also a plane wave, $\psi_i\psi_f^* = \exp(iqr)$, satisfying

$$q = p_f - p_i.$$

The relationship of these wave vectors to the scattering angle is shown in Fig. 1.10. For elastic scattering, the wavenumber q is expressed in terms of the scattering angle θ by the simple formula

$$q = 2p\sin(\theta/2). \tag{1.1}$$

The amplitude for scattering at the corresponding angle θ is directly proportional to the Fourier component of the potential field felt by the projectile at wavenumber q. In a finite system, the Fourier transform varies smoothly with q and thus the scattering will be a smooth function of angle rather than sharply peaked as in scattering from crystals.

Our first example of diffractive scattering in a finite system is the scattering of electrons from the nucleus ^{208}Pb. The differential cross section is shown in Fig. 1.11 as a function of the scattering angle. The three curves show the cross section for elastic scattering and inelastic scattering to the first and second excited states of the nucleus ^{208}Pb. The bombarding energy of the electrons is 500 MeV, giving them a wavelength corresponding to an angular wavenumber $p = E/\hbar c \sim 2.5$ fm^{-1}. The cross section falls off very steeply with scattering angle, reflecting the fact that the Coulomb potential is a smooth function with a steeply falling Fourier transform. However, the angular distribution also shows a mild oscillatory behavior, which is traceable to the finite size of the charge distribution.

It is easy to make a qualitative numerical connection between the size of the nucleus, the wavelength of the electron, and the

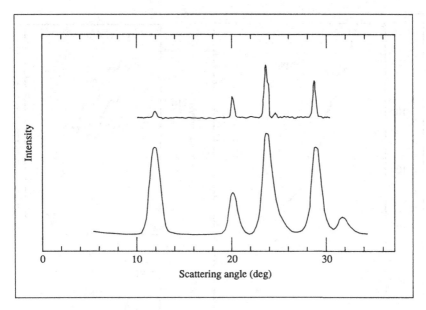

Fig. 1.9. X-ray and neutron diffraction patterns for magnetite at room temperature (from Kittel (1968)).

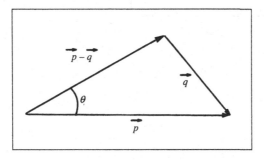

Fig. 1.10. Inelastic scattering of a fast particle with angular wavenumber $\vec{p}$. Relation between the scattering angle θ and momentum transfer $\hbar\vec{q}$.

period in the diffraction pattern. The observed angular distribution shows an oscillation with successive maxima spaced by about 11°. From eq. (1.1), this corresponds to a change in the angular wavenumber by about $q \sim 0.5$ fm^{-1}. An extended object with a well-defined surface at a distance R from the center has an oscillatory Fourier transform whose extrema are approximately separated by an angular wavenumber q satisfying

$$qR \approx \pi. \tag{1.2}$$

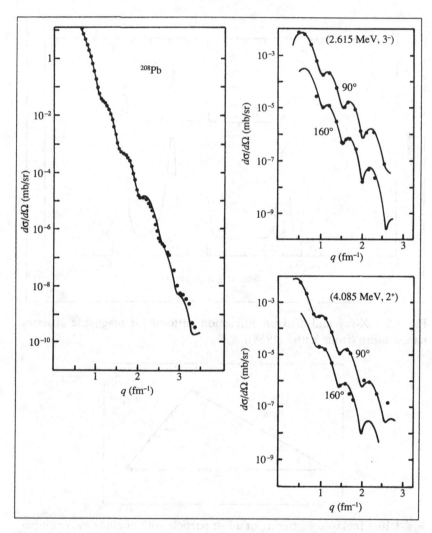

Fig. 1.11. Elastic and inelastic electron scattering differential cross section associated with the ground state, and with the lowest 3^- and 2^+ states of ^{208}Pb, as a function of momentum transfer. The elastic data is from Frois et al. (1977); the inelastic scattering data has been taken from Heisenberg (1981).

This implies, for small angle scattering, that the maxima in the scattering cross section are separated by

$$\Delta\theta = \frac{\pi}{pR}. \tag{1.3}$$

Thus we infer directly from the electron scattering data that the nucleus ^{208}Pb has a radius of about 6.5 fm. This argument is, of course, very rough; the actual theory of the scattering incorporates deviations of the electron wave function from the plane waves, and allows one to infer accurately the complete shape of the charge distribution in the nucleus.

In Fig. 1.12 we show a similar experiment using protons instead of electrons as projectiles. In the region shown, only the nuclear interaction is important. The diffractive structure is more pronounced in this case for two reasons. First, the nuclear interaction is short-range and its Fourier transform does not fall off as rapidly as that of the Coulomb interaction. Second, the projectile is absorbed in the interior of the nucleus, and its wave function cannot be treated as a plane wave any more. The interior absorption, in effect, concentrates the wave physics on the outer part of the nucleus and emphasizes the role of the nuclear surface. Again, the diffraction pattern relates directly to the wavelength of the projectile and the size of the nucleus. The calculation inferring the nuclear size from this data is carried out in the figure caption.

We next give an example of electron scattering from the atomic physics domain. Experiments are usually done with electrons of rather low energy. This makes the experiments more difficult to interpret, because the electron wave functions are strongly distorted. The example is electron scattering on Xe atoms, measured with electrons of 1 eV energy. The angular distribution is shown in Fig. 1.13. This shows a very strong diffraction structure, with a strong maximum at forward scattering angles, a minimum near 30°, and a secondary maximum near 90°. In this example, the bombarding electron has far too low an energy to be treated as a plane wave. The diffractive structure can be understood only in terms of partial wave decomposition of the projectile wave function or detailed numerical models of the potential and the resulting wave function distortion. In any case, if one tries the same estimates as discussed previously, one obtains $q \approx 0.5$ Å^{-1} and $R \approx 4$ Å, which is somewhat larger than the xenon radius.

1.3 Energy transfer in inelastic scattering

We now examine inelastic scattering more closely from the point of view of the energy transferred to the system. The most detailed

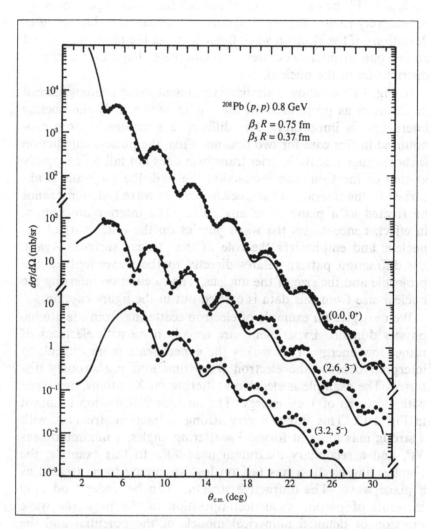

Fig. 1.12. Angular distribution of elastic and inelastic proton scattering for the ^{208}Pb ground state and excited states at 2.6 and 3.2 MeV are given. The data was taken from Blanpied et al. (1978). In this case, the energy of the proton was $E_p = 800$ MeV and its angular wavenumber k is obtained from the relativistic formula $(E_p + m_p c^2)^2 = (m_p c^2)^2 + (\hbar p c)^2$, giving $p = 7.4$ fm^{-1}. The diffraction peaks in the data are separated by about 3.5°, consistent with eq. (1.3) for a radius $R \approx 6.8$ fm.

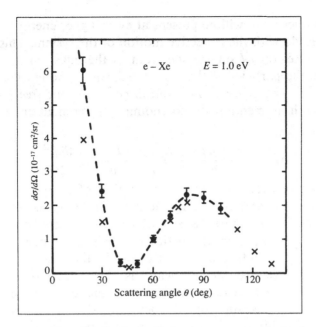

Fig. 1.13. Angular distribution of electrons scattered elastically from xenon at a bombarding energy of 1 eV (from Weyhreter et al. (1988)).

measurements are in nuclear physics, and our examples come mostly from this area. The momentum transfer and associated wave number q are well-defined quantities when we deal with weakly interacting projectiles, and we can discuss the distribution of inelastic scattering as a function of energy loss, keeping the momentum transfer fixed. A schematic energy distribution function is shown in Fig. 1.14. Starting from zero energy loss, wc first see a peak associated with elastic scattering. This is shown as Region I in the figure; experimentally the energy resolution would smear the peak over some interval of energy. Going up in energy a little bit, of the order of an MeV, we come to Region II which contains discrete energy levels that can be excited and studied individually. The density of these levels increases very sharply with energy, and at some point individual levels can no longer be resolved. Nevertheless bumps may still be present in the spectrum. One of these is marked as Region III in the figure. This is the region of the giant resonances, which occur at excitations of the order of ten to twenty MeV. They can be excited by inelastic scattering as well as by photon absorption. When the momentum transfer $\hbar q$ is large, another peak will be present at even higher energy that has nothing to do with the collective motion of the system. This peak, shown as Region IV on the spectrum, is the quasielastic scattering. Here the particles of the system behave as though they were almost free. The energy conservation condition for free particles starting with momentum P and ending with momentum $(P + \hbar q)$ reads

$$\Delta E = \frac{(P + \hbar q)^2}{2m} - \frac{P^2}{2m} = \frac{\hbar P \cdot q}{m} + \frac{(\hbar q)^2}{2m}$$

If we average over the relative orientation of P and q the first term drops out and the energy loss is $(\hbar q)^2/2m$, as for a free particle recoil. The peak is spread out by the contribution of the first term, and thus the width of the peak depends on the momentum distribution of the particles. In the case of nucleons in a nucleus, the characteristic width is of the order of $2\hbar P_f q/m$, where $P_f \sim 270$ MeV/c is the Fermi momentum. Examples of inelastic electron scattering on nuclei are shown in Figs. 1.15 and 1.16. The first figure shows electron scattering from a ^{208}Pb target. The peaks from discrete states are visible at low excitation energies, extending up to about 5 MeV. Above that, the main structure in the spectrum is due to the giant resonances, located near 12 MeV, 14 MeV, and 22 MeV. The momentum transfer in this reaction is not large enough to produce a quasielastic peak. In the next example, Fig. 1.16, the momentum transfer is much higher and most of the cross section goes to the quasielastic part of the spectrum.

In condensed matter physics, inelastic electron scattering is also useful to measure properties of excitation modes. In fact, the technique is commonly referred to as EELS, for electron energy loss spectroscopy. In a bulk metal, the most important mode is the plasmon, and this is the main feature of th energy loss spectroscopy in the eV region. An example is shown in Fig. 1.17. There are a number of peaks, for two reasons. Bulk plasmons and surface plasmons have different frequencies, and both are excited. Also, the plasmon couples so strongly to electrons that several plasmons can be excited by a single electron.

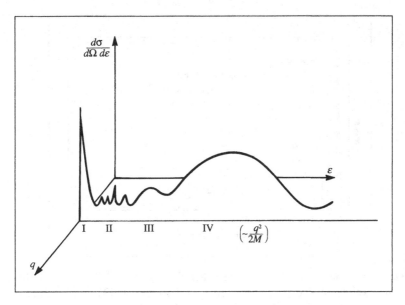

Fig. 1.14. A schematic double differential cross section for electron scattering at a particular momentum transfer. The cross section exhibits: (I) elastic scattering, (II) excitation of discrete levels, (III) giant resonances and (IV) the quasielastic peak.

1.4 Inelastic scattering with strongly interacting projectiles

Inelastic scattering with projectiles like protons and alpha particles provides useful data on nuclear excitations, including the giant resonances. Nuclear projectiles have advantages and disadvantages with respect to electrons as probes of the nucleus. Nucleon scattering is more difficult to analyze because the wave function distortion effects are large. Also, the interaction between projectile and target is not known as well for the nuclear forces as for electromagnetic interaction.

On the other hand, there are compensating advantages to choosing nuclear projectiles to excite the nucleus. The interaction is dominated by the nuclear force, which can excite different modes than the electromagnetic force. The nuclear force is charge symmetric, that is, the same for neutrons and protons. The scattering of alpha particles will therefore strongly favor excitations in which the neutrons and protons move together. Part of the nuclear force involves charged fields, such as the pion field. Excitations mediated

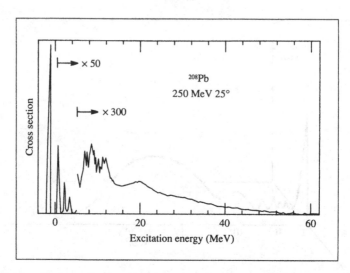

Fig. 1.15. Electron scattering data for the nucleus ^{208}Pb as a function of the excitation energy E_{ex} (Sasao and Torizuka (1977)). The electron energy is 250 MeV and the scattering angle is 25°, which corresponds to a momentum transfer of 100 MeV/c. The strongest peak is the elastic scattering at $E = 0$. In the region below 5 MeV one sees discrete lines corresponding to excited states of the ^{208}Pb nucleus. At higher energy, the individual states merge together. The prominent peaks in the region 10–15 MeV are the giant resonances. Note the change in scale between the different energy regions.

by the pion field can thus change the charge of the nucleus. The response of the nucleus to charge-changing fields is very important in understanding the beta decay process, which also changes the charge of a nucleus. Thus indirect information about beta decay is obtained from nucleon scattering experiments.

Two geometric modes

Another advantage of nuclear scattering is that the strong absorption makes the diffraction quite sensitive to the characteristics of the excitation in the surface of the nucleus. We will illustrate this by contrasting the diffraction from two important nuclear giant resonance modes. One of these modes, to be discussed in detail in Chap. 6, is the giant quadrupole resonance of which the physical motion of the particles is shown in Fig. 1.18. The spherical shape is distorted into an ellipsoid, and the motion produces alternately

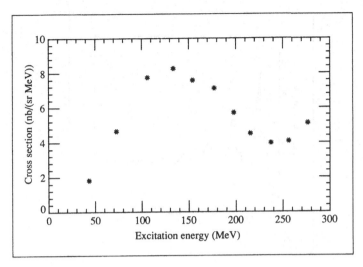

Fig. 1.16. Electron scattering from ^{40}Ca in the quasielastic region of excitation energy (Deady et al. (1986)). The incident electron energy is 375 MeV and the scattering angle is 90°. Using the relativistic relation between energy and momentum leads to a momentum transfer of the order of 500 MeV/c. The quasielastic peak is seen at about 120 MeV excitation energy. The rising cross section at the highest energy is attributed to internal excitations in the nucleon (cf. Fig. 1.8).

elongated and squashed shapes of the nucleus. The other simple mode we mention, which will be discussed in detail in Chap. 7, is called the breathing mode or giant monopole resonance. In it the nucleus vibrates in the radial direction, oscillating between states of relative compression and rarefaction. This motion is illustrated in Fig. 1.19. These modes are induced by the projectile wave function, which becomes diffracted in the process. To see what the diffractive effects are, we look at the geometry of the projectile wave, depicted in Fig. 1.20 for a strongly interacting projectile. Here the projectile wave envelopes the nucleus, interacting with it mainly at the periphery. Portions of the wave that penetrate into the nucleus give rise to very strong interactions, which would not allow the projectile to emerge again intact. To understand the diffractive effects associated with these modes, let us imagine a projectile going by a nucleus and interacting with it. The detected projectile could only have passed by the periphery of the nucleus because, if the projectile entered the interior, it would interact very

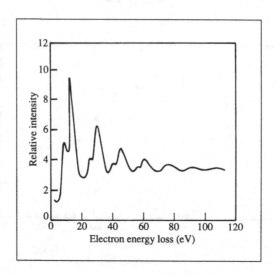

Fig. 1.17.   Energy loss spectra for electrons reflected from a film of aluminum, for a primary electron energy of 2020 eV. The 11 loss peaks observed in Al are made up of combinations of 10.3 and 15.3 eV losses, where the 10.3 eV loss is due to surface plasmons and the 15.3 eV loss is due to volume plasmons. After C.J. Powell and J.B. Swan, *Phys. Rev.* 115, 869 (1959); 116, 81 (1959); cf. Kittel (1968) p. 234.

strongly and would not emerge again without a very large energy loss.

The projectile can induce some motion of the target at its periphery. According to the laws of quantum mechanics, the scattered wave acquires a phase depending on the phase of the motion which the nuclear surface acquires in the excitation. Thus, for the monopole mode, where all the radial motion has the same sign, the scattered waves move forward in phase with each other. This produces a diffraction pattern with a maximum at the center, i.e. for scattering into the forward direction. This is illustrated in Fig. 1.21, where two wavelets from different sides of the target come with a constructive interference in the forward direction. This effect is analogous to the Poisson spot seen in optics for diffraction of light around spheres; the phenomenon is a bright spot seen on the beam axis in the forward shadow of the sphere.

Turning now to the quadrupole mode, the projectile excitation induces part of the periphery to move outward and part inward, distorting the equator into an ellipse. With different phases at different points of the equator, the scattered wave no longer has

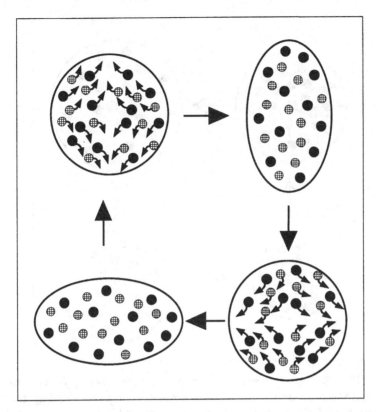

Fig. 1.18. Schematic representation of the giant quadrupole resonance in nuclei. This mode is a shape vibration, in which both the shape and the distribution of nucleons change. In this vibration both protons and neutrons move in phase distorting the system from a spherical shape into an ellipsoidal shape and back. This vibration can be excited efficiently by projectiles such as alpha particles, which interact with the surface of the target, and have the same interaction for neutrons and protons.

constructive interference for forward scattering. This is depicted in Fig. 1.22. From simple geometry the angle of the maximum peaking can be estimated. It is approximately $\theta \approx L/pR$, where L is the multipolarity of the mode. As before, p is the angular wavenumber of the projectile and R is the radius of the nucleus. Thus, the qualitative signature of each geometric mode is a strong peak at a scattering angle that depends on the type of vibration. For quadrupole vibrations, $\theta \approx 2/pR$. Data illustrating the excitation of monopole and quadrupole modes of the nucleus by alpha particles are shown in Fig. 1.23. This displays the cross section for

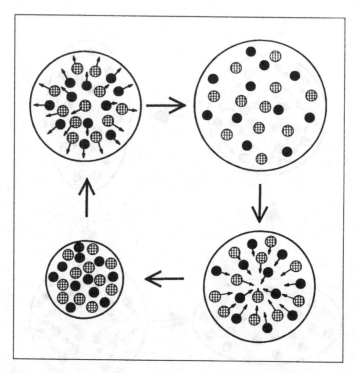

Fig. 1.19. Schematic representation of the giant monopole vibration.
In it both proton and neutrons move radially in phase, leading to an
expansion and a contraction of the nucleus. This is the reason why
this vibration is called a breathing mode. The compressibility of nuclear
matter may be inferred from the observed frequency of the vibration (see
Chap. 7).

exciting the giant resonances in the nucleus ^{208}Pb as a function of
scattering angle. The curves drawn along the data points are the
theoretical distributions predicted with projectile wave functions
calculated in an optical model, for which there exists a number of
computer programs. The quadrupole mode has an angular distribution peaking at 4°. The simple formula above yields $\theta \approx 3.8°$,
taking the reduced wavenumber as $p = \sqrt{2M_\alpha E} \approx 4.3$ fm^{-1} and
the radius as $R \approx 7$ fm. At larger angles, the diffraction is oscillatory and dominated in character by the interference of the
projectile wave passing opposite sides of the nucleus. For the
monopole excitation, the data does not extend all the way to zero
degrees, due to technical experimental problems. However, one

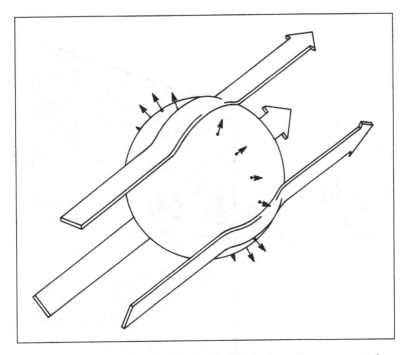

Fig. 1.20. A projectile wave, depicted with the broad arrows, envelops a nucleus and induces the transverse motion of the nucleus, shown with the small arrows. This motion resolves into a superposition of the monopole and the quadrupole normal modes.

can see in the optical model predictions and in the trend of the data toward smaller angles that the major peaking is as expected.

1.5 Spin excitations

The modes we have looked at up to now, namely the dipole, monopole, and quadrupole giant vibrations, are all physical motions with clear analogies in the vibrations of ordinary macroscopic bodies. In quantum mechanics the spin of the particles can also participate in collective excitations, and we include them in our study. In the physics of bulk condensed matter, these modes are known as magnons. The spin of the electron interacts with low energy neutrons in a way favorable to the excitation of magnons. In nuclear physics, protons in the energy range of several hundreds of MeV interact fairly strongly with the spin degrees of freedom

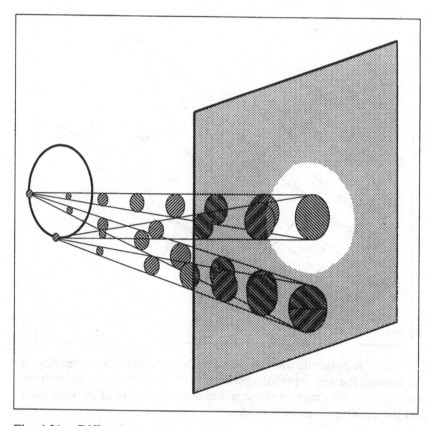

Fig. 1.21. Diffraction pattern associate with monopole vibrations. The interaction of the wave describing the motion of the projectile with the nucleons of the surface of the target alters the form of the wave. After the interaction each small area of the nuclear surface acts as the origin of a spreading wavelet. The phase of the wavelet depends on the motion of the originating surface. Because the motion of the surface in a monopole vibration is completely symmetric, the wavelets originating from the nuclear surface all have the same phase. Along the beam axis, where the distance to all regions on the periphery of the sphere is equal, the wavelets arrive in phase and interfere constructively. Consequently the differential cross section displays a maximum at $\Theta = 0°$.

of the target nucleons, exciting spin modes. At the same time that the spin is excited, charge can be transferred to the nucleus, changing a target neutron into a proton or vice versa. Of course, to conserve total charge the projectile charge would change in the opposite way. In spite of these complications of spin and charge

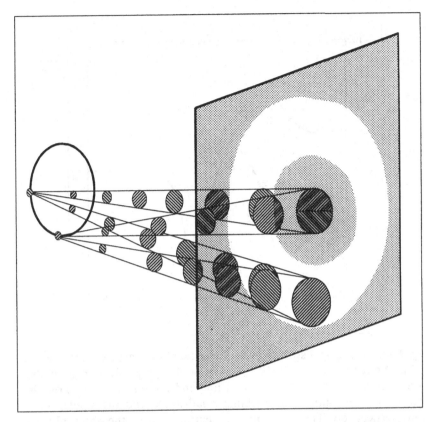

Fig. 1.22. Diffraction pattern associated with the quadrupole vibrations. Wavelets originating from the elongating sides of the nucleus are initially 180° out of phase with the wavelets originating in the contracting sides. As a consequence, the interference in the center of the plane perpendicular to the beam axis is destructive. There are other points in the plane, however, where the distance to an elongating and to a contracting region just differ by half a wavelength, leading to constructive interference. Because the nucleus can have any orientation, this coherence leads to a ring centered on the axis of the beam and thus to a maximum of the inelastic differential cross section at a finite angle.

states, these scattering reactions display the same diffractive effects seen in the geometric modes. The simplest possible spin vibration, which is uniform over the entire surface of the nucleus, gives rise to a diffraction pattern peaking at zero degrees, just as the giant monopole does. The charge-exchange spin vibration, called the giant Gamow–Teller resonance because of its relation

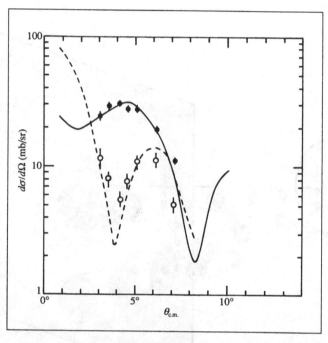

Fig. 1.23.   Angular distribution for inelastic alpha particle scattering on
^{208}Pb (Youngblood et al. (1977)). The alpha projectile energy in this
measurement is 100 MeV. The solid line and the dashed lines show the
theoretical distributions for exciting quadrupole and monopole vibrations, respectively. The data points are extracted from the observed cross
sections at $E = 11.0$ MeV (solid circles) and 13.7 MeV (open circles).

to the spin-flip beta decay process, is the most prominent feature of the charge exchange reaction of protons on heavy nuclei.
An example is shown in Fig. 1.24.

1.6   Excitation by heavy ions

Heavy ion reactions, reactions in which a nucleus is bombarded
by another large nucleus, also provide a powerful probe of nuclear
excitations. Heavy ions have several characteristics that make them
quite distinct from lighter projectiles. The Coulomb and to some
extent the nuclear forces are much stronger in heavy ions. This is
illustrated by the excitation of giant resonances. Fig. 1.25 shows
the inelastic excitation function for projectiles of ^{17}O bombarding
a ^{208}Pb target. One sees the narrow states at low excitation and the

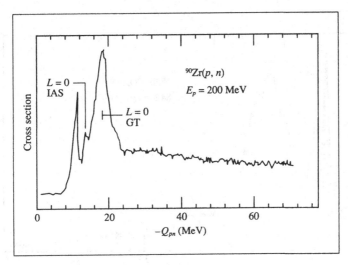

Fig. 1.24.   Neutron spectra from (p, n) reactions on ^{90}Zr at 200 MeV (from Gaarde (1981)).

giant dipole as the dominant feature of the spectrum at 14 MeV excitation. The dipole peak is relatively larger in this reaction than with alpha-induced reactions because the eight charges in the oxygen projectile produce a larger Coulomb field. Coulomb excitation cross sections are proportional to the square of the charge of the projectile nucleus.

Another characteristic of heavy ion projectiles useful in the study of nuclear vibrations is that the interaction is entirely peripheral, even more than with proton, neutron, or alpha projectiles. At closer distances of approach, the projectile and target interact very strongly, absorbing much energy and destroying the quantum diffractive effects. An economical description of these processes can be carried out describing the relative motion of the ions by a complex potential, where the imaginary potential represents in some average way all the processes depopulating the initial beam. This model is known as the optical model, in analogy with scattering of light in a dispersive medium. As a consequence of the strong surface absorption, one needs only the optical model wave functions at fairly large separations, where they can be calculated rather accurately. A third feature of heavy ion reactions is that the wavelength associated with the relative motion is quite short, typically two orders of magnitude smaller than the nuclear dimensions. Classical physics can then be used to describe the

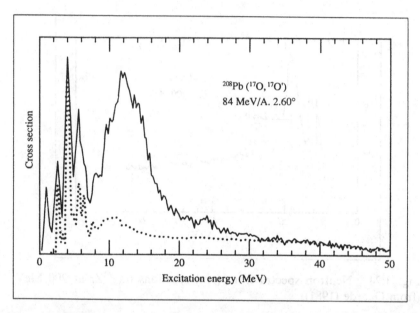

Fig. 1.25. Spectra from inelastic scattering of ^{17}O on ^{208}Pb (from Bertrand, Beene and Horen (1988)). The solid and dotted lines show cross sections at beam energies of 22 and 84 MeV/nucleon, respectively.

trajectory of the ion. On the other hand, the classical behavior means that diffraction patterns will be weak, and there will not be much of a signature of the spatial distribution of the modes from diffractive peaking of the angular distribution. Another limitation in the use of heavy ion reactions to study collective states is the technical problem that excitations of the projectile can become confused with those of the target. However, this problem can be alleviated by choosing suitable projectiles.

Fig. 1.26 shows an elastic scattering angular distribution. It is plotted as a ratio to the Rutherford cross section, which varies with angle as $\sigma_R \sim 1/\sin^4(\theta/2)$. The large-impact parameter collisions give only small deflection to the projectile, and thus for small angles the cross section is close to Rutherford. This may be seen to be the case in Fig. 1.26 for angles up to about 4°. Beyond that, the cross section falls steeply. This is due to the strong interaction and absorption of the projectile when they come close together. It is also clear from the smooth behavior of the cross section that diffraction effects are rather small, but still visible.

Inelastic scattering of heavy ion projectiles is very simple to

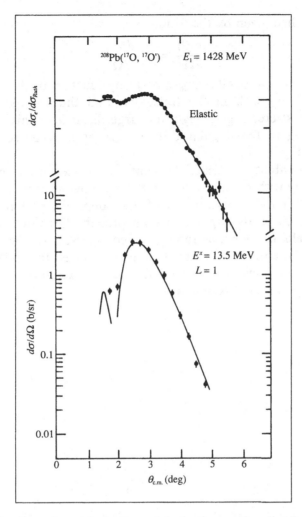

Fig. 1.26. Experimental elastic and inelastic angular distributions for the reaction ^{208}Pb $(^{17}O,^{17}O')$ at a bombarding energy of 1428 MeV associated with the ground state and the giant dipole resonance of ^{208}Pb (from Barrette et al. (1988)).

describe when the excitation is weak, as for example with a giant resonance. Then the trajectory of the heavy ion is practically the same as for elastic scattering. The excitation process can be treated by time-dependent perturbation theory, which yields a probability P_{0n} to excite some state n along a given trajectory. The cross section is then given by the formula

$$\frac{d\sigma(\theta)}{d\Omega} = P_{0n}\frac{d\sigma_{el}}{d\Omega}.$$

Of course, the probability P_{0n} is a strong function of the distance of closest approach in the trajectory, and thus of the impact parameter, decreasing strongly for large distances. This explains the peaking in the angular distribution for inelastic scattering, shown in Fig. 1.26.

The probability of exciting giant resonances increases in a marked way with bombarding energy, as is obvious from Fig. 1.26. This is a simple consequence of time-dependent perturbation theory, which yields larger transition probabilities when the perturbing field is changing rapidly in comparison to the rate of vibration. In keeping with this fact, it is expected that relativistic heavy ion reactions may permit the study of excitations built out of two or more giant resonances.

2

Basic concepts

Atomic nuclei and the electrons in metal clusters may be described as systems of fermions of approximately uniform density. When the systems are spherical, their radii R scale with the number of particles N as

$$R = r_0 N^{1/3} \; .$$

The quantity r_0 is the Wigner–Seitz radius of the sphere associated with the volume occupied by each fermion. In the case of atomic clusters, r_0 is conventionally expressed in atomic units, $r_0 = a_0 r_s$, where $a_0 = \hbar/m_e e^2 = 0.529$ Å is the Bohr radius and m_e the mass of the electron. For metal clusters r_s varies between 2 and 6, depending on the element. The associated interior density n_0 has the order of magnitude of 10^{22} cm^{-3}. In the nuclear case, the radius parameter is $r_0 \approx 1.2$ fm, and the associated saturation density[*] is about $\rho_0 \approx 0.16$ fm$^{-3} \approx 10^{38}$ cm^{-3}. With a large number of particles, the behavior of both nuclei and the electrons in metal clusters will follow the dynamics of continuum mechanics. In this chapter we will mainly review the classical mechanics needed to describe this behavior.

2.1   Vibrations of continuous systems

We now review the theory of vibrations in classical systems. The simplest case is a single degree of freedom. For later reference

---

[*] We will follow the usual conventions in the literature and use the symbols n and ρ to denote electron densities and nuclear densities, respectively.

we remind the reader of the fundamental equations. Almost all small-amplitude vibrations are harmonic, and the dynamics follows from Newton's equation with a linear restoring force. The linear restoring force arises from a Taylor series expansion of the potential energy function about its minimum, $V(x) = V(0) + \frac{1}{2}kx^2+...$ Newton's acceleration equation then reads

$$m\frac{d^2x}{dt^2} = -kx,$$

where x is the displacement of the particle and m is its mass. The motion obeying this equation is sinusoidal, $x = x_0 \sin \omega t$, with an angular frequency

$$\omega = \sqrt{\frac{k}{m}}. \tag{2.1}$$

We shall now jump immediately to continuous systems, representing the limit of large numbers of particles. The system has some equilibrium configuration, and the motions will be described by the displacement of the medium from the equilibrium. The displacement can depend on position of course, so the degrees of freedom are in a continuous vector field. We will call $\vec{u}(r)$ the displacement at point r. We may also want to consider the possibility of several different kinds of media, for example protons and neutrons in a nucleus. Then we would define a separate field for each kind of particle. To construct the dynamics we must have an energy function and be able to write down the kinetic and potential energies. The kinetic energy is easily expressed in terms of the velocity field of the motion, $d\vec{u}/dt$, the mass of a particle m, and the particle density ρ:

$$T = \frac{1}{2} \int d^3r \sum_i m_i \rho_i \frac{d\vec{u}_i}{dt} \cdot \frac{d\vec{u}_i}{dt}. \tag{2.2}$$

Here we have indicated with the index i the possibility of several kinds of particles.

The potential energy function is much more subtle. Its specific form is dictated by physical considerations. For example, it can only depend on differences or gradients of the displacements, because the energy is independent of the absolute position of the system. A commonly assumed form is the potential energy associated with some two-particle interaction, which we may write, with the help of the particle density ρ, as

$$\mathcal{V} = \frac{1}{2} \sum_i \int d^3r d^3r' \rho_i(r) v(r - r') \rho_i(r') \quad . \tag{2.3}$$

Introducing the potential field

$$V(r) = \int d^3r' v(r - r') \rho(r') \quad , \tag{2.4}$$

associated with the motion of individual particles, one can also write the potential energy function as

$$\mathcal{V} = \frac{1}{2} \sum_i \int d^3r \rho_i(r) V_i(r) \quad . \tag{2.5}$$

Another common assumed form for $\mathcal{V}$ is a function of the densities alone,

$$\mathcal{V} = \int d^3r f[\rho_i(r)] \quad . \tag{2.6}$$

This form requires that the interactions be short-ranged, since different spatial regions contribute independently. Each of the above expressions will prove useful in different situations, as we shall see below.

Let us suppose we have some potential function that has been expressed in terms of the fields $\vec{u}$. How do we find the small amplitude vibrations of the system? The standard procedure starts from the Lagrangian, the difference of the kinetic and potential energy functions. From a variational principle one extracts a partial differential equation for the fields $\vec{u}$. The desired solutions must also satisfy certain boundary conditions.

The Lagrangian formulation is too involved for our purposes, so we shall use another method that is physically more straight-forward. This is the Rayleigh–Ritz variational principle. In this method, any form of the $\vec{u}$ field is allowed as a guess, and one extracts an associated frequency. This will be the actual vibrational frequency if the frequency is a minimum with respect to variations in $\vec{u}$. Very often one has a good guess for the field $\vec{u}$ and then the resulting frequency is close to the actual one. The formula for the Rayleigh–Ritz variational frequency is very simple.

To derive it, suppose that there is some collective coordinate $a(t)$ describing the oscillations of the system. One can write the time dependent displacement fields as

$$\vec{u}(r, t) = \vec{u}_0(r) a(t) \quad , \tag{2.7}$$

with the normalization

$$a(t)|_{max} = 1. \tag{2.8}$$

Assuming furthermore that the motion is harmonic, at an angular frequency ω to be determined later, one obtains

$$\dot{a}(t)|_{max} = \omega \ .$$

One can then write down the kinetic energy at the maximum using $d\vec{u}(r,t)/dt|_{max} = \omega\vec{u}_0(r)$,

$$T|_{max} = \frac{1}{2}\omega^2 m \int d^3r \rho(r)\vec{u}_0 \cdot \vec{u}_0 \ . \tag{2.9}$$

It is convenient to define a generalized inertia,

$$I = m \int d^3r \rho(r)|\vec{u}_0(r)|^2 \ , \tag{2.10}$$

so the maximum kinetic energy may be expressed as

$$T|_{max} = \frac{1}{2}I\omega^2.$$

In harmonic motion, the energy is transferred back and forth between kinetic and potential, and the maximum kinetic energy is equal to the change in potential energy between equilibrium and maximum displacement. Of course, this assumes that the potential energy can be accurately described as a quadratic form in the displacement, which is part of our overall assumption of harmonic motion. The restoring force constant C for the collective motion is related to the maximum potential energy by

$$C = \frac{\partial^2 \mathscr{V}}{\partial a^2} = 2(\mathscr{V}(u_0)|_{max} - \mathscr{V}(0)) \ . \tag{2.11}$$

Setting the maximum potential energy equal to the maximum kinetic energy, we obtain the final equation

$$\omega^2 = \frac{C}{I} \ . \tag{2.12}$$

This is Rayleigh's variational principle. It looks just the same as the formula for the frequency of oscillation of a mass on a spring. In effect, we have reduced the problem to a single degree of freedom by fiat, when we defined the field $\vec{u}$ to represent the vibrational motion. In general, eq. (2.12) gives an upper bound on the frequency of the lowest vibrational modes.

As a first general application, let us consider a potential energy function of the form of eq. (2.3). The $a(t)$-dependence of the potential comes from the displaced densities,

$$\rho(\vec{r} + \vec{u}) = \rho_0(r) + a(t)\vec{u}_0(r) \cdot \vec{\nabla}\rho_0(r) \ . \tag{2.13}$$

It is convenient to denote the spatial dependence of the density oscillation as

$$\delta\rho(r) = (\rho(\vec{r} + \vec{u})_{max} - \rho_0) = \vec{u}_0 \cdot \vec{\nabla}\rho_0 \ . \tag{2.14}$$

Then to second order in $a(t)$ the potential energy is given by

$$\mathscr{V}(u) = \mathscr{V}(0) + \frac{a^2(t)}{2} \int d^3r d^3r' \delta\rho(r) v(r - r') \delta\rho(r') \ . \tag{2.15}$$

We did not include a linear term in $a(t)$ because we always assume that the system is in equilibrium at $a = 0$. The quantity

$$\mathscr{V}(0) = \frac{1}{2} \int d^3r d^3r' \rho_0(r) v(r - r') \rho_0(r)$$

is the equilibrium potential energy. We then determine C from eq. (2.11) as

$$C = \int d^3r d^3r' \delta\rho(r) v(r - r') \delta\rho(r') \ . \tag{2.16}$$

It will sometimes be useful to introduce the oscillating spatial dependence of the potential field as we did for the density. We define the oscillating potential field $\delta V(r)$ putting the time-varying density in eq. (2.4), and extracting the oscillating part of $V(r)$. This is simply

$$\delta V(r) = \int d^3r' \delta\rho(r') v(r - r') \ . \tag{2.17}$$

When we come to the quantum mechanical theory, the δV will be the same as the transition potential. In terms of δV, the restoring force constant in Rayleigh's principle is

$$C = \int d^3r \delta\rho\delta V \ . \tag{2.18}$$

Example of dipole oscillations

We illustrate the above discussion with a model of the nuclear giant dipole resonance proposed by Goldhaber and Teller (1948). In this model, neutrons and protons move in opposite directions. The simplest assumption we could make about the field shape is that the motion is the same everywhere. Thus the proton displacement would be some constant, say the unit vector in the z direction, $\vec{u}_{0p} = \vec{u}_0 = \hat{z}$, and the neutron displacement would be the opposite, $\vec{u}_{0n} = -\vec{u}_0$. The inertia associated with this field is very simple: since there is no spatial dependence, the integral over density is just the total mass,

$$I = m \int d^3r \left(\rho_p \hat{z} \cdot \hat{z} + \rho_n(-\hat{z}) \cdot (-\hat{z}) \right) = mZ + mN = mA \quad . \quad (2.19)$$

In the above equation m is the nucleon mass, and Z, N, and A are the proton, neutron, and nucleon numbers, respectively.

Goldhaber and Teller assumed that the potential energy depends quadratically on the difference between the neutron and proton densities. This is plausible from several points of view. The binding energies of nuclei contain a symmetry term, proportional to the square of the difference between neutron and proton numbers. One can argue that the nuclear forces tend to keep an equal number of neutrons and protons, and the energy cost of a deviation should go quadratically with the deviation. Thus we take our potential energy function to be

$$\mathscr{V} = \int d^3r v_{sym} \left[\rho_n(r) - \rho_p(r) \right] \quad , \quad (2.20)$$

with v_{sym} an even function of its argument. Let us assume that the neutron and proton densities are the same in equilibrium. The displaced densities are then given by

$$\rho_p(\vec{r} + \vec{u}) = \frac{1}{2}(\rho_0 + a(t)\vec{u}_0 \cdot \vec{\nabla}\rho_0) \quad ,$$

$$\rho_n(\vec{r} - \vec{u}) = \frac{1}{2}(\rho_0 - a(t)\vec{u}_0 \cdot \vec{\nabla}\rho_0) \quad . \quad (2.21)$$

We insert this in eq. (2.20) and expand to obtain for the numerator function of the Rayleigh principle

$$C = \frac{\partial^2 \mathscr{V}}{\partial A^2} = \int d^3r \frac{\partial^2 v_{sym}}{\partial \rho^2} (\nabla_z \rho_0)^2 \quad (2.22)$$

$$= \frac{4\pi}{3} \int_0^\infty r^2 dr \frac{\partial^2 v_{sym}}{\partial \rho^2} \left(\frac{d\rho_0}{dr} \right)^2 \quad .$$

In the last step we used the fact that $\partial r/\partial z = \cos\theta$ and assumed that the density ρ_0 is spherically symmetric. Making use of the relations eq. (2.12) and (2.19), we obtain the final expression for the frequency

$$\omega^2 = \frac{4\pi}{3mA} \int_0^\infty r^2 dr \frac{\partial^2 v_{sym}}{\partial \rho^2} \left(\frac{d\rho_0}{dr}\right)^2 . \tag{2.23}$$

To proceed to a numerical result, we would need to know the actual functions v_{sym} and $\rho_0(r)$. We can deduce some things about the qualitative behavior just from the functional form of the above potential energy function, however. Note that the derivative of ρ appears in the integral, so the important contributions to the integral come from the surface where the density is varying rapidly. Nuclei have surface properties that are rather independent of the size of the nucleus. Thus the surface thickness and the symmetry energy coefficient can be taken to be independent of A. The surface position however, varies as $A^{1/3}$ as in an incompressible drop. Consequently, the main dependence on A in eq. (2.22) will come from the factor r^2 in the integral, making C roughly proportional to $A^{2/3}$. Thus, the vibrational frequency will depend on mass number A roughly as

$$\omega \sim \sqrt{A^{2/3}/A} = A^{-1/6} .$$

This model correctly describes the mass dependence in very light nuclei, but not in heavier nuclei. We shall discuss the reasons in detail in Chap. 5; the main point is that the field $\vec{u}$ will no longer be a constant when we are dealing with a larger system. In fact, it takes more energy to move a neutron through the nuclear surface, than to move it from one place to another in the interior.

Example of quadrupole oscillations

Another example of a possible mode of oscillation of a sphere is a quadrupolar shape vibration. The displacement field $u_0(r)$ for this case is particularly simple if the sphere is incompressible. This is achieved by having a field of the form $\nabla r^L P_L(\cos\theta)$, with P_L a Legendre function of angular momentum L. The incompressibility then follows from the relation $\vec{\nabla} \cdot \vec{u} = \nabla^2 r^L P_L(\cos\theta) = 0$. Any flow generated by a potential field such as this is also irrotational (i.e. $\vec{\nabla} \times \vec{u} = 0$; cf. also Chap. 6.). For the quadrupole mode $L = 2$, we can write the displacement field as

$$\vec{u} \sim \vec{\nabla}(z^2 - 1/2(x^2 + y^2)) = (-x, -y, 2z) . \tag{2.24}$$

The field (2.24) distorts the system by stretching it in the z-direction and compressing it in the transverse direction (cf. App. B). Overall, there is no compression if the amplitude of the displacement is small. It is convenient to define a displacement field that produces no compression even for large amplitudes. One way to do this is by scaling the Cartesian coordinates:

$$x' = x \exp(-\epsilon)$$

$$y' = y \exp(-\epsilon) \qquad (2.25)$$

$$z' = z \exp(2\epsilon) \ .$$

Here ϵ is a dimensionless measure of the amplitude of the quadrupole distortion. It is obvious from the above equation that volumes are preserved by the transformation. To lowest order, the displaced density can be expressed in terms of the equilibrium density as

$$\rho(\vec{r} + \vec{u}) = \rho(x(1 - \epsilon), y(1 - \epsilon), z(1 + 2\epsilon)) \ .$$

Let us take ϵ to be a collective coordinate and see how the oscillation frequency would be calculated. The first task is to compute the inertia, which requires the velocity field $\dot{u}$:

$$\dot{u} = \vec{u}_0(r)\dot{\epsilon} = (-x, -y, 2z)\dot{\epsilon} \ .$$

Inserting $u_0(r)$ in eq. (2.10) we find

$$I_{irr,Q}^{\epsilon} = m \int d^3r \rho(r)(4z^2 + x^2 + y^2) \ .$$

Here the labels *irr*, Q and ϵ indicate that the motion is irrotational, quadrupolar, and the inertia is with respect to the coordinate ϵ. At the spherical shape this can be expressed in terms of the mass of the system and mean square radius $< r^2 > \equiv \int d^3r \rho(r) r^2 / \int d^3r \rho(r)$ as

$$I_{irr,Q}^{\epsilon} = 2Nm < r^2 > \approx \frac{6}{5} NmR^2 \ . \qquad (2.26)$$

Here N is the number of particles and m is the mass of each particle. In the approximate equality above, we further assumed that the system is spherical with uniform density and radius R, implying $< r^2 >= \frac{3}{5}R^2$.

We next need the potential energy of the system as a function of deformation. There are two well-known extremes for the possible behavior of a classical system governed by short-range interactions. One is fluid behavior. Then the energy does not depend on the distortion in the interior, since no compression has occurred. However, the energy associated with the surface will change because the amount of surface area is increased by the distortion. This gives rise to the surface tension, which provides a restoring force to the spherical shape. The dependence of the area on the deformation is given to second order in ϵ by the formula (cf. eq. (6.6))

$$\mathscr{A} = 4\pi R^2 (1 + \frac{8}{5}\epsilon^2 + ...) \; .$$

The surface energy is expressed in terms of the surface tension coefficient γ as $\mathscr{V} = \mathscr{A}\gamma$, giving an effective liquid drop restoring force constant

$$C^\epsilon_{liq,Q} = \partial^2\mathscr{V}/\partial\epsilon^2 = 64\pi R^2\gamma/5 \; . \qquad (2.27)$$

Then the frequency of quadrupolar oscillations in an incompressible liquid drop is obtained from the inertia, eq. (2.26), and the above spring constant as

$$\omega_{liq,Q} = \sqrt{\frac{C^\epsilon_{liq,Q}}{I^\epsilon_{irr,Q}}} = \sqrt{\frac{32\pi\gamma}{3mN}} \; . \qquad (2.28)$$

In the nuclear case we identify m with the nucleon mass M; $M_p = 938.3$ MeV/c^2 and $M_n = 939.6$ MeV/c^2, and N with the atomic mass number A. The surface tension has an empirical value given approximately by $\gamma \approx 1$ MeV/fm^2. The above equation leads to an excitation energy $\hbar\omega_{liq,Q} \approx 37/A^{1/2}$ MeV. As an example, we mention the spherical nucleus ^{120}Sn, for which this formula gives $\hbar\omega_{liq,Q} \approx 3.4$ MeV, to be compared with the empirical value of 1.32 MeV for the lowest quadrupolar excitation. The formula gives values which are systematically too high, although the right order of magnitude is obtained. We shall see in Chap. 6 compelling arguments to reject the simple liquid drop model for describing such modes. It will emerge that while the restoring force C_{liq} is indeed correct on average, the inertia $I^\epsilon_{irr,Q}$ of eq. (2.26) is much too small (cf. Sec. 6.4 and Fig. 6.8). In fact, the motion in the low-lying nuclear vibrations has considerable vorticity.

If the system is solid rather than liquid, the potential energy function is completely different. The energy density in a homogeneous, isotropic solid is conventionally expressed as

$$\mathscr{V} = \frac{\lambda}{2}(\nabla \cdot u)^2 + \mu \sum_{i,j} \left(\frac{\nabla_i u_j + \nabla_j u_i}{2}\right)^2 , \qquad (2.29)$$

where λ and μ are the Lamé coefficients of elasticity (Landau and Lifshitz (1970)). For incompressible fields, the parameter λ plays no role and the energy depends only on μ, the shear modulus of elasticity. In an infinite system, the vibrations are characterized by two sound velocities. The longitudinal sound has a velocity given by

$$v_l^2 = \frac{\lambda + 2\mu}{m\rho} , \qquad (2.30)$$

and the transverse sound velocity is given by

$$v_t^2 = \frac{\mu}{m\rho} .$$

Returning to the oscillations of a spherical solid, the strain energy associated with the quadrupole field of eq. (2.25) is

$$\mathscr{V} = \int d^3r \mu(6\epsilon^2) = \frac{4\pi}{3} R^3 6\epsilon^2$$

and the spring constant is

$$C_{el,Q} = \frac{48\pi \mu R^3}{3} .$$

Finally, with the inertia eq. (2.26) the elastic vibrational frequency is

$$\omega_{el,Q} = \sqrt{\frac{6\mu}{m\rho <r^2>}} . \qquad (2.31)$$

We shall see in Sec. 6.3 (cf. eq. (6.15)) that this relation provides an accurate treatment of the giant quadrupole resonance in nuclei. A slightly lower frequency can be obtained by including a compressional component in the displacement field. According to the variation principle, this would give a better approximation to the actual motion of the system in the lowest resonance. The detailed classical theory was given by Lamb (1882).

Let us illustrate the above formula with an example of a metal cluster, the sodium cluster Na_{20}. We may express eq. (2.31) in terms of the radius of the sphere R and the transverse sound velocity v_t as

$$\omega_{el,Q}^2 \approx \frac{10\mu}{m\rho R^2} = \frac{10v_t^2}{R^2} \ .$$

Taking parameter values $v_t = 1000$ m/s and $R = 5.7$ Å, the angular frequency is $\omega_{el} = 5.3 \times 10^{12}$ s^{-1}. The excitation energy of a vibrational quantum in this mode is $\hbar\omega_{el,Q} = 3.6$ meV. Although no direct information exists on these vibrations, it will be seen in Chap. 9 that the frequencies are important to understand the damping rate of electronic excitations in metal clusters.

2.2 Resonance formulas

Up to now we have just considered the free oscillations of a harmonic system. It will be important to understand the coupling to outside forces, and the effect of the presence of dissipative mechanism leading to damping. All this fits into the common mathematical structure of the resonant formulas and, in this section, we will briefly review it.

The concept of resonance appears everywhere in the dynamics of physical systems. All isolated multi-particle systems have natural vibrational frequencies, and the resonance phenomenon describes how these modes couple to the outside environment. Remarkably, the mathematical structure of the resonance formulas is universal, not depending on the details of the physical system at all. Of course, the parameters in the formulas can be specified only from a knowledge of the system's dynamics.

We first derive the Lorentz resonance formula, using the simplest possible physical realization of a vibrating system, a mass connected to a spring. To this is added a frictional force proportional to velocity, $F_{fric} = g\,dx/dt$, where x is the displacement of the mass and g is a constant. We then obtain the equation of motion from Newtonian mechanics:

$$m\frac{d^2x}{dt^2} + g\frac{dx}{dt} + kx = F_{ext} \tag{2.32}$$

where F_{ext} is an external force. Let us suppose the external force is sinusoidal with an angular frequency ω,

$$F_{ext} = F_0 e^{-i\omega t} \ . \tag{2.33}$$

In classical physics all quantities are real, and we can interpret the above equation by taking its real part only,

$$F_{ext} = \text{Re}(F_0)\cos\omega t + \text{Im}(F_0)\sin\omega t \ .$$

The main advantage of the complex notation is that both sine and cosine dependences can be expressed with one simple function, the complex exponential. In quantum mechanics of course the complex arithmetic is essential. Then eq. (2.32) is solved by the function

$$x(t) = \frac{F_0 e^{-i\omega t}}{-m\omega^2 - ig\omega + k} \ .$$

It is also convenient to write this in terms of $\omega_0 = \sqrt{k/m}$,

$$x(t) = \frac{F_0 e^{-i\omega t}/m}{(-\omega^2 - i\gamma\omega + \omega_0^2)} \ . \qquad (2.34)$$

Here we have defined a damping constant $\gamma = g/m$. To get the physical quantities from this complex equation, we next extract the real part. Assuming F_0 real, this has the form

$$\text{Re}(x(t)) = \frac{F_0}{m}\text{Re}(\Pi)\cos\omega t + \text{Im}(\Pi)\sin\omega t \ ,$$

where

$$\text{Re}(\Pi) = \frac{\omega_0^2 - \omega^2}{(\omega_0^2 - \omega^2)^2 + \gamma^2\omega^2}$$

$$\text{Im}(\Pi) = \frac{\gamma\omega}{(\omega_0^2 - \omega^2)^2 + \gamma^2\omega^2} \ .$$

The rate at which energy is expended on the motion is proportional to the product of the velocity

$$\text{Re}(\dot{x}(t)) = \frac{F_0\omega}{m}[\text{Im}(\Pi)\cos\omega t - \text{Re}(\Pi)\sin\omega t]$$

and the force

$$\text{Re}(F_{ext}) = F_0\cos\omega t \ .$$

When averaged over time, only the $\text{Im}(\Pi)$ term contributes; the power dissipation rate is

$$P = \frac{F_0^2}{m}\frac{\gamma\omega^2}{(\omega_0^2 - \omega^2)^2 + \gamma^2\omega^2} \ . \qquad (2.35)$$

This is the famous Lorentz formula for resonances. It describes

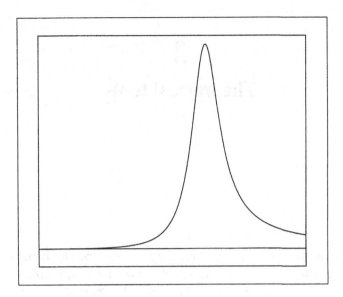

Fig. 2.1. Lorentzian resonance profile, eq. (2.35).

the coupling of an individual vibrational mode to external fields in the presence of damping. The function is shown in Fig. (2.1). It might seem very restrictive to demand that the frictional force be linear in the velocities, but this is the most common situation for small amplitude motion. When the system is coupled to a continuum, as for example an oscillating charge damped by the radiation of electromagnetic energy, the linear damping is exact.

In quantum mechanics a similar formula arises when the Schroedinger equation is solved in the presence of a time-varying external field. This is known as the Breit–Wigner formula in scattering theory. For example, the probability of producing a resonant state by scattering a particle from a system depends on the energy of the particle E as

$$P \sim \frac{1}{(E - E_r)^2 + (\Gamma/2)^2} \, . \tag{2.36}$$

This function is the same as the Lorentzian function in the vicinity of the resonance when $\gamma << \omega_0$. The Lorentz formula provides a quite accurate description of the photoabsorption resonance observed in heavy nuclei (cf. Chap. 5).

3

Theoretical tools

It might be thought that the best way to describe the behavior of a many-particle system is to develop the complete quantum mechanical wave function and then evaluate any required observables. This naive way of proceeding is successful for small molecules, but the complexity of the wave function grows enormously as the number of particles increases, so there is no hope of following this path for any but the smallest systems. Instead, one applies formal methods that just pick out the aspects of the behavior that are physically of interest, and applies approximation schemes that emphasize these aspects. In order to do this, there are a number of formal tools and identities that make the analysis simpler. In this chapter, we shall explain some of the most important, leading up to the RPA theory of collective oscillations, which we develop in the next chapter.

3.1   Operators

What are the physical observables relevant to vibrational dynamics? Fundamental are the density and current operators, whose expectation values tell us where the particles are and what their momentum is. For a particle in a potential well the matrix element of these two operators is given by the expressions

$$< i|\rho(\vec{r})|j >= \phi_i^*(\vec{r})\phi_j(\vec{r})$$

$$< i|\vec{j}(\vec{r})|j >= \frac{\hbar}{2im}(-(\nabla\phi_i^*)\phi_j + \phi_i^*\nabla\phi_j)$$

46

where ϕ_i and ϕ_j are the initial and final wave functions. The definition of these operators in a many-particle wave function is more complicated to write down; one must integrate over all particles in the wave function but one. Specifically, we have

$$< i|\rho(\vec{r})|j> = \sum_k \int \psi_i^*(r_1...r_n)\psi_j(r_1...r_n) \prod_{l \neq k} dr_l \qquad (3.1)$$

and

$$< i|\vec{j}(\vec{r})|j> = \frac{\hbar}{2im} \sum_k \int \left(-(\nabla_k \psi_i^*)\psi_j + \psi_i^* \nabla_k \psi_j \right)/2im \prod_{l \neq k} dr_l \ .$$
$$(3.2)$$

For actual calculations with many-particle wave functions, one often uses a basis of independent single-particle wave functions. It is then convenient to define these operators in terms of creation and annihilation operators of the single-particle states in Fock space. However, we shall not need such detailed methods for our purposes here.

Vibrations may be described with time-dependent wave functions. The expectation of the density operator in a time-dependent wave function corresponds directly to the classical picture discussed in Chap. 2 of a physically moving system whose density changes in time, and gives a clear way to see the relation between quantum and classical dynamics. If the oscillation is undamped, we can write the expectation of the density operator in the time-dependent state as

$$< \rho(\vec{r}) > = \rho_0 + \delta\rho \, \cos(\omega(t - t_0))$$

where ρ_0 is the static density distribution, and $\delta\rho$, called the **transition density**, is a function of $\vec{r}$ alone.

A common situation will be where the motion produces no compression of the medium. The density still fluctuates because the boundary position changes in time. In this case, if the amplitude of vibration is small, the amount of the density fluctuation will be proportional to the amplitude of the motion in the direction perpendicular to the surface. The functional form of the transition density in this case is

$$\delta\rho = \vec{u}_0 \cdot \nabla\rho_0 \qquad (3.3)$$

where $\vec{u}_0$ is the displacement field of the vibration. This intuitively plausible formula, which exactly corresponds to the classical oscillating density in eq. (2.14), will emerge from the operator relations developed in the next section.

We can define a transition current in the same way as the transition density. It will have the time dependence

$$< \vec{j} >= \delta \vec{j}(r) \sin(\omega(t - t_0)) \ .$$

For the limiting case of incompressible motion with a displacement field $\vec{u}$, the transition current will have the form

$$\delta \vec{j} = \vec{u}_0 \rho_0 \ .$$

To describe vibrations such as the nuclear giant dipole, where neutrons oscillate against protons, or spin excitations in metal clusters, we need to consider density operators for the isospin and the spin, respectively. For example, the spin density has three components for the three Cartesian directions. The z-component is the difference in density of particles with spin up and down, and the other components are defined with the Pauli spin matrices accordingly. In a one-particle wave function of the type $(\phi(r, \uparrow), \phi(r, \downarrow))$ the spin density has matrix elements,

$$< i|\sigma_z(r)|j >= |\phi(r, \uparrow)|^2 - |\phi(r, \downarrow)|^2 \qquad (3.4)$$

$$< i|\sigma_x(r)|j >= \phi_i^*(r, \uparrow)\phi_j(r, \downarrow) + \phi_i^*(r, \downarrow)\phi_j(r, \uparrow)$$

and

$$< i|\sigma_y(r)|j >= -i\phi_i^*(r, \uparrow)\phi_j(r, \downarrow) + i\phi_i^*(r, \downarrow)\phi_j(r, \uparrow) \ .$$

This can be easily generalized to many-particle wave functions as we did with the ordinary density, and an analogous spin current $\vec{j}_\sigma$ can also be defined.

To describe the nuclear giant dipole state, we will use the z component of the isospin density. Mathematically, this is very similar to the spin density, with the spin-up component representing a proton and the spin-down component representing a neutron. The density for the isospin operator τ_z is just the difference of neutron and proton densities, as in eq. (3.4) above. The other isospin operators, namely τ_x and τ_y, change neutrons into protons and vice versa, and so produce densities which one does not usually associate with classical vibrations. However, we will see later that these operators are important for describing processes in which the charge of the system changes.

We have indicated the form of the operator expectation values in a time-dependent description, but one more often uses a time-independent representation to describe quantum process. To make the connection, one must be able to express the initial wave function in terms of the energy eigenfunctions. These have a simple time dependence,

$$\psi_i(t) = \exp(-iE_it/\hbar)\psi_i(0) \ .$$

Our vibrating quantum state must contain at least two different eigenstates in order for physical quantities to vary with time. Let us write such a state as

$$\psi(t) = \exp(-iE_1t/\hbar)\psi_1 + \exp(-iE_2t/\hbar)\psi_2 \ .$$

The expectation of the density operator in this state has a time-varying part that is proportional to

$$\psi_1^*\psi_2 e^{-i(E_2-E_1)t/\hbar} + \psi_1\psi_2^* e^{i(E_2-E_1)t/\hbar} \ .$$

The two factors in the above equation differ by only a phase; if we take ψ_1 and ψ_2 to be real the two terms combine to

$$2\psi_1\psi_2 \cos(\omega t) \ .$$

Thus we see that the transition density is proportional to the matrix element of the density operator. The precise normalization will be discussed later.

3.2 Sum rules

There are some very important operator identities that restrict the possible matrix elements in a physical system. The most important is the commutator of the Hamiltonian with the density operator, which is given by

$$\frac{i}{\hbar}[H, \rho(r)] = \nabla \cdot \vec{j}(r) \ . \tag{3.5}$$

This is proven in quantum mechanics textbooks[*] and we will just accept it as given. To apply this relation, we take matrix elements of the operators between states of interest. On the right hand side, the matrix element is the divergence of some transition current. On the left hand side, the matrix element simplifies if the two states are eigenstates of the Hamiltonian. Then the operator H

---

[*] See, for example, Sect. 19 of Landau and Lifshitz (1974).

just gives the energy of the state times the same state, and the matrix element is proportional to a transition density. The result is

$$\frac{1}{\hbar}(E_i - E_j)\delta\rho = \nabla \cdot \delta\vec{j} \ . \tag{3.6}$$

Thus if we happen to know the form of the transition current, the shape and magnitude of the transition density will be readily calculable without looking at the wave functions in any more detail. Conversely, if the quantum mechanical approximation scheme yields matrix elements that do not satisfy this identity, it shows a severe weakness in the approximation.

The above relation is nothing more than the conservation law for particles. In classical physics, this law states that the divergence of the current is equal to the time derivative of the density. It is also true as an operator relation in quantum mechanics, and with matrix elements between eigenstates the time derivative is replaced by the energy difference.

We next examine the effect of a sudden external field on a system that starts out in equilibrium. In a time-dependent picture, the effect of the field will be to induce a current. To see how this comes about, we define the field $V(r) = F(r)\delta(t)$ with a spatial dependence $F(r)$ and a δ-function time dependence that acts at $t = 0$. Note that $F(r)$ has the dimension of energy×time, while $V(r)$ has energy dimensions. Then from the time-dependent Schrödinger equation we determine the effect of the field on the wave function. Because of the δ-function time dependence the perturbation is easy to evaluate immediately after $t = 0$. After the perturbation and to first order in V, the wave function is given by

$$\psi(+\epsilon) = \psi_0 + iF(r)\psi_0/\hbar$$

where we have indicated the time just after the impulse as $+\epsilon$ and the initial wave function as ψ_0. We now evaluate the expectation of the current operator. The derivatives in the current operator act on ψ_0 as well as F. If there is no current in the initial states, the terms with derivatives on ψ_0 cancel. What remains is

$$< \vec{j}(r) >_{+\epsilon} = < \psi(+\epsilon)|\vec{j}(r)|\psi(+\epsilon) >$$

$$= \frac{\hbar}{m} \sum_i \int \psi_0^* \nabla F(r_i)\psi_0\delta(r - r_i) \prod_{j \neq i} dr_j$$

$$= \rho(r)\nabla F(r)/m \ . \tag{3.7}$$

This simple result is exactly what one would expect from classical physics. If we start with a system in equilibrium and apply an impulsive potential field, the particles are accelerated and their momentum is given by the impulse associated with the force. This relates to the above equation with the momentum density at r equal to $m < \vec{j}(r) >$ and the impulse per particle given by $\nabla F(r)$.

The above identity becomes a sum rule if we expand the state $\psi(+\epsilon)$ in terms of the energy eigenstates and their amplitudes. This is given by

$$\psi(+\epsilon) = \psi_0 + \frac{1}{\hbar} \sum_i \psi_i < i|F|0 > \quad .$$

Then eq. (3.7) implies that

$$2 \sum_i < 0|F|i >< i|\vec{j}(r)|0 >= \hbar \rho_0(r) \nabla F(r) \quad . \tag{3.8}$$

We can get a sum rule for the transition density by taking the divergence of this equation, and replacing $\nabla \cdot \vec{j}$ with eq. (3.7). This yields

$$\sum_i < 0|\rho(r)|i >< i|F|0 > (E_0 - E_i) = \frac{\hbar^2}{2m} \nabla \cdot \rho_0(r) \nabla F(r) \quad . \tag{3.9}$$

If the system has collective oscillations, it may happen that $< i|F|0 >$ is small for all states but the collective excitation. In that case only one term in the sum would be important, and we could use eq. (3.8) to determine the form of the transition density. The result is

$$\delta \rho = \frac{\hbar^2}{2m} \frac{\nabla \cdot \rho_0 \nabla F}{(E_0 - E_i) < 0|F|i >} \quad .$$

This functional form of the transition density with F taken as pure multipole fields was introduced by Tassie (1956),

$$\delta \rho \sim \nabla \cdot \rho_0 \nabla r^L Y_{LM} \quad \text{(Tassie model)} . \tag{3.10}$$

We next derive a general energy-weighted sum rule for matrix elements from eq. (3.9), multiplying it by $F(r)$ and integrating over r. On the left hand side, note that $\int d^3r < 0|\rho(r)F(r)|i >=< i|F|0 >$. On the right hand side, we integrate by parts once to get the identity

$$\sum_i |<0|F|i>|^2(E_i - E_0) = \frac{\hbar^2}{2m} \int d^3r |\nabla F|^2 \rho_0$$

(energy-weighted sum rule).   (3.11)

Like the other identities, the energy-weighted sum rule has a very simply classical interpretation. We used the field V to give the particles a momentum ∇F. On the average, the particles started at rest so their average energy after the sudden impulse is $|\nabla F|^2/2m$. Thus the right hand side of eq. (3.11) is just the average energy given to the system. It is independent of the interactions because the energy is absorbed before the system is disturbed from equilibrium. On the left hand side, $|<0|F|i>|^2/\hbar^2$ is the probability of the state ψ_i in the perturbed system. Thus the left hand side is the excitation energy of the system, calculated from the decomposition into energy eigenstates. If there is a single collective state, this sum rule can be used to calculate its strength in terms of its excitation energy. Then combining this with eq. (3.8), we could determine the transition density with its normalization.

The energy-weighted sum rule is most often used in finite systems for multipole fields, $F(\vec{r}) = r^L Y_{LM}(\hat{r})$. The right hand side can then be simplified algebraically if the system is spherically symmetric. The result is

$$\sum_i |<0|r^L Y_{LM}(\hat{r})|i>|^2(E_i - E_0) = \frac{L(2L+1)\hbar^2}{8\pi m} \int d^3r\, r^{2L-2} \rho_0 .$$

(3.12)

Note that the sum rule is independent of M, the magnetic quantum number of the multipole field. In nuclear physics, the matrix elements of multipole charge fields arise in the description of photon transitions and inelastic scattering of charged projectiles. These are often expressed as the $B(EL)$, defined for states of arbitrary angular momentum J as

$$B(EL; J_i \to J_f) = \frac{1}{2J_i+1} \sum_{M_i,M_f,M} <J_iM_i|er^L Y_{LM}|J_fM_f>^2 .$$

In terms of the $B(EL)$, the energy-weighted sum rule for a $J_f = 0$

state reads

$$\sum_f B(EL; L_f \to 0)(E_f - E_i) = \frac{L(2L + 1)\hbar^2}{8\pi m} \int d^3r r^{2L-2} \rho_0 \ .$$

3.3 TRK sum rule and the oscillator strength

The most important application of the sum rule identities is to the case with a constant force field, i.e. F having a constant gradient. The electric field from a photon is of this form in the dipole approximation, which is valid when the size of the system is small compared to the wavelength of the photon. Inserting $F = z$ in eq. (3.11), the integral simplifies because $|\nabla F|^2 = 1$ and the integral over the density is just the number of particles,

$$\sum_i | < 0|z|i > |^2 (E_i - E_0) = \frac{\hbar^2 N}{2m} \quad \text{(TRK sum rule)}. \qquad (3.13)$$

This is known as the Thomas–Reiche–Kuhn sum rule (TRK). Just this combination of matrix element and energy appears in the expression for the photon absorption cross section, as will be demonstrated in the next section.

It is useful to define a dimensionless quantity f which is the measure of the fraction of the sum rule taken by a single transition. The definition is

$$f_{0i} = \frac{2m}{\hbar^2} < 0|z|i >^2 (E_i - E_0) \ . \qquad (3.14)$$

Then the TRK sum rule for a single particle is one, and in general

$$\sum_i f_{0i} = N \quad (f \text{ sum}). \qquad (3.15)$$

3.4 Photon cross section

In this section we present the formulas for photon absorption and emission in the dipole approximation, and give a derivation of the photoabsorption sum rule. The dipole photon emission rate of a system decaying from a state i to a state 0 is derived in quantum mechanics textbooks[†], and is given by

$$\Gamma_\gamma = \frac{4}{3}e^2 <i|\vec{r}|0> \cdot <0|\vec{r}|i> k_\gamma^3 \qquad (3.16)$$

where $k_\gamma = (E_i - E_0)/\hbar c$ is the reduced wavenumber of the photon. As a consequence of the decay, the state has a natural width Γ_γ.

The state also appears as a resonance in the scattering of photons from the ground state. The resonant cross section is given by a geometric factor π/k_γ^2 multiplied by a statistical factor and a resonance function of the Lorentzian shape. If the resonance is narrow compared to the photon energy, the formula is

$$\sigma = \frac{S}{2}\frac{\pi}{k_\gamma^2}\frac{\Gamma_\gamma^2}{(E - E_0)^2 + (\Gamma_\gamma/2)^2}.$$

In the statistical factor $S/2$, S is the degeneracy of the excited states divided by the degeneracy of the ground state. Thus, for the common situation where f, the lower state, has spin zero and the excited state i has spin 1, the statistical factor is $S = 3$. The factor $1/2$ in $S/2$ is due to the two-fold degeneracy of the polarization in photon states. Integrating the above cross section over energy, we find that an isolated resonance contributes

$$\int_{resonance} \sigma dE = S\pi^2\Gamma_\gamma/k_\gamma^2 = \frac{2\pi^2 e^2\hbar}{mc}f_{0i}. \qquad (3.17)$$

In the last equality, we used eq. (3.12) and (3.13), assuming a spherically symmetric ground state and a dipolar excited state with $S = 3$. According to eq. (3.17), the integrated photon absorption cross section depends only on the oscillator strength, and not otherwise on the properties of the excitation.

It is now a simple matter to apply the TRK or f-sum rule to get a sum rule for the photon absorption cross section. Including all states in the integral of eq. (3.17) gives a total integrated photon cross section

$$\int \sigma dE = \frac{2\pi^2 e^2\hbar N}{mc}. \qquad (3.18)$$

It may seem from this discussion that the validity of the sum rule requires that the states only have their natural width and are well separated, but in fact the sum rule is much more general. The conditions for the validity are:

[†] For example, eq. (20.39) of Merzbacher (1961).

- that the dipole approximation is valid, i.e. the photon wavelength is large compared to the size of the system;

- the equation of continuity is satisfied by the currents and the densities;

- the only charges in the system are the N particles of mass m.

We now present another derivation of the sum rule which makes the generality of the formula more evident. We follow the physical interpretation of the TRK sum rule as the energy given to the system by an external impulse. In this derivation, consider the interaction of the system with a narrow wave packet of photons, as shown in Fig. 3.1. This will satisfy the condition of an impulsive external force. Let us assume the wave moves in the x-direction and depends only on $x - ct$. We will take the electric field to be in the z-direction and strongly peaked as a function of $x - ct$. For computational convenience, we use a Gaussian shape for the electric field,

$$\vec{\mathscr{E}}(t) = \hat{z}\mathscr{E}_0 \exp(-(x - ct)^2/2x_0^2) \ .$$

Of course, in the end nothing should depend on the shape of the field and in particular on the width parameter x_0 that was introduced.

The force acting on each charged particle is $e\vec{\mathscr{E}}$. Consequently, the impulse $\int dt\, e\vec{\mathscr{E}}$ from the field gives a momentum p to each particle according to

$$\vec{p} = \hat{z}\int dt\, e\mathscr{E}_0 \exp(-(x - ct)^2/2x_0^2) = e\hat{z}\mathscr{E}_0\sqrt{2\pi}x_0/c \ .$$

The total energy given to the system is

$$E_{abs} = N\frac{p^2}{2m} = \frac{2\pi Ne^2\mathscr{E}_0^2 x_0^2}{2mc^2} \ . \tag{3.19}$$

We now examine the energy transfer from a different point of view, namely that the electromagnetic wave may be represented as a superposition of photons, which interact independently according to their cross section. If the number of photons per unit area and in a frequency interval $d\omega$ is defined as $N(\omega)d\omega$, the energy associated with those photons is $\hbar\omega N(\omega)d\omega$ and the average energy absorbed by the system can be expressed in terms of the photon absorption cross section σ as

$$E_{abs} = \int d\omega\, \hbar\omega N(\omega)\sigma \ .$$

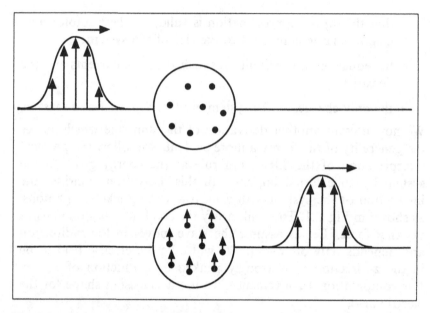

Fig. 3.1. A narrow electromagnetic wave passes a system and impulsively accelerates its charged particles.

All that is left to do is to figure out the photon density in the above Gaussian electromagnetic field. To get the frequency dependence we make a Fourier transform of the field,

$$\tilde{\mathscr{E}}(\omega) = \int dt \mathscr{E}(t) e^{i\omega t} = \frac{\mathscr{E}_0 \sqrt{2\pi} x_0}{c} \exp(-(\omega x_0/c)^2/2) \ .$$

The energy density is proportional to the modulus squared of the field,

$$dE = C e^{-(\omega x_0/c)^2} d\omega$$

where C is some constant. Integrating this equation we obtain the total energy per unit area

$$E = C \sqrt{\frac{\pi}{4} \frac{c}{x_0}} \ .$$

Another way to determine the total energy per unit area is from classical electromagnetism

$$\int dx (\mathscr{E}^2 + \mathscr{B}^2)/8\pi = \frac{1}{4\pi} \int dx \mathscr{E}^2 = \sqrt{\pi} \mathscr{E}_0^2 x_0/4\pi \ .$$

From this we may determine the normalization constant $C =$

$x_0^2 \mathscr{E}_0^2 / 2\pi c$, and then the equivalent photon energy density. Integrating the photon energy density over frequencies from zero to infinity, the properly normalized photon energy density is found to be

$$\hbar \omega N(\omega) = \frac{\mathscr{E}_0^2 x_0^2}{2\pi c} e^{-(\omega x_0/c)^2} .$$

According to the impulse approximation, x_0 should be taken small enough so that the variation of $\hbar \omega N(\omega)$ over the range of frequencies that the system responds to can be neglected. Then we can write

$$E_{abs} = \int d\omega \hbar \omega N(\omega)\sigma = \frac{\mathscr{E}_0^2 x_0^2}{2\pi c} \int d\omega \sigma .$$

Comparing with eq. (3.19) we obtain the desired sum rule,

$$\int dE\sigma = \frac{2\pi^2 \hbar e^2 N}{mc} .$$

Notice that the details of how we constructed the field cancelled out.

In chemistry, one often uses a formula that relates the absorption directly to the sum rule fraction f. As mentioned in Chap. 1, the absorption is generally parameterized in terms of the **molar extinction coefficient** ϵ according to the equation relating the incident intensity I_0 to the transmitted intensity I,

$$I = I_0 10^{-\epsilon \rho_M x_{cm}} .$$

Here ρ_M is the density of absorbing centers in moles per liter and x_{cm} is the thickness of the absorbing layer in cm units. The relation between the cross section σ and ϵ is then

$$\sigma = \epsilon \frac{\rho_M}{\rho} \frac{x_{cm}}{x} \log_e 10 .$$

In the above units $\rho_M/\rho = 1000/N_0$, where N_0 is Avogadro's number. We also express the formula for f, eq. (3.16), in terms of an integral over wave number instead of an integral over energy. The formula then reads

$$f = \frac{mc^2}{\pi e^2} \int \sigma dv.$$

Combining this with the above, we obtain

$$f = \frac{1000mc^2}{\pi e^2 N_0} \log_e 10 \int \epsilon dv$$

$$= 4.32 \times 10^{-9} \int \epsilon dv$$

valid for ϵ defined in the [l-M^{-1}cm^{-1}] units and v in [cm^{-1}] units.

3.5 Spin sum rules

A completely different kind of sum rule is useful in analyzing the response to operators that depend on spin or isospin. These sum rules start from operator commutator identities. For example, the spin operators σ_x and σ_y satisfy the identity,

$$[\sigma_x, \sigma_y] = i\sigma_z \quad .$$

We can look at a particular matrix element of the identity, namely the expectation value in some state. Let us at the same time write out the two terms in the commutator and insert a complete set of states between the σ operators in the products. This is

$$i < i|\sigma_z|i> =$$

$$\sum_f < i|\sigma_x|f> < f|\sigma_y|i> - < i|\sigma_y|f> < f|\sigma_x|i> \quad .$$

This can be rewritten as

$$i < i|\sigma_z|i> = \frac{1}{2i}\sum_f |<i|\sigma_x + i\sigma_y|f>|^2 - \sum_f |<i|\sigma_x - i\sigma_y|f>|^2 \quad .$$

Thus we obtain a relation between squares of matrix elements and a static expectation value. The particular example here is not so interesting, because the state i usually has spherical symmetry. This implies that the expectation value is zero, and that the two terms on the right hand side are equal and opposite.

A more interesting application is to the spin–isospin excitations of nuclei. We defer a detailed discussion to Chap. 8.

3.6 Polarizability sum

Under the influence of an external electric field a system of charged fermions will polarize. Assuming the field to be uniform and pointing along the z-direction one obtains

$$V = -e\mathscr{E}z$$

for the interaction energy of the field and the induced dipole moment. Because of this interaction the wave function ψ_0 describing the system in the absence of the electric field will be modified. In first order perturbation theory one obtains

$$\psi = \psi_0 + \sum_i \frac{< i|V|0 >}{E_0 - E_i}\psi_i$$

$$= \psi_0 + e\mathscr{E}\sum_i \frac{< i|z|0 >}{E_i - E_0}\psi_i \ .$$

The static polarizability α of a system is the ratio of its dipole moment to the electric field that induces the polarization. Consequently

$$\alpha = \frac{\text{dipole moment}}{\text{electric field}} = 2e^2 \sum_i \frac{(z)_{0i}^2}{E_i - E_0} \ . \tag{3.20}$$

If α is measured or if we have another way to calculate it, this equation provides a useful constraint on the excitation spectrum, which we shall develop in more detail in the chapter on dipole excitations. A classical correspondence can be made for the above formula if we express the squared matrix element in terms of f. One then obtains

$$\alpha = e^2 \sum_i \frac{f_{0i}}{m\omega_{0i}^2} = e^2 \sum_i \frac{f_{0i}}{k_{0i}} \ . \tag{3.21}$$

In the last equality, we have defined an effective spring constant for each transition with $\omega^2 = k/m$. The polarizability is the same as that for classical particles on springs weighted with a factor f for each spring.

The polarizability concept can be generalized in several ways. We may define polarizabilities for arbitrary fields in the same way as we did for the dipole polarizability. For example, replacing ez by the quadrupole field $z^2 - y^2 - x^2$ in eq. (3.20) gives the quadrupole polarizability. This can be estimated in simple models and provides a helpful guide to the quadrupole excitation properties.

If the transition strength for an operator is concentrated in a single state, its frequency may be calculated from the polarizability together with the EWSR. With a single state contributing, the summations may be dropped in eq. (3.11) and (3.20) and the ratio of the sum rules is

$$\omega_{ad}^2 = \frac{\int d^3r |\nabla V|^2 \rho_0/m}{\alpha_V} . \tag{3.22}$$

Of course, in a physical situation the strength will be distributed over many states, and then this formula gives a certain weighted average of the transition energy. In this average, low frequencies are weighted heavily (cf. eq. (3.20)), corresponding to an adiabatic limit. This is denoted with the subscript *ad* in the above formula. The adiabatic estimate will prove to be very useful in discussing the dipole oscillation of electrons in Chap. 5.

The polarizability concept can also be generalized to higher order terms in the external field. This is relevant to situations where multiple excitations are possible. Nuclear experimental techniques at present are just beginning to allow multiple dipole excitations to be measured. In atomic cluster physics, it is relatively easy to induce multiple transitions with intense laser beams. However, this topic is beyond the scope of this book.

Another generalization of the polarizability is to time-dependent external fields. Let us consider an external field of the form $V(r)\cos\omega t$. Then the expectation value of V has a time-varying part, proportional to $\cos\omega t$, and its coefficient is given by

$$\alpha_V(\omega) = \sum_i <i|V|0>^2 \left(\frac{1}{E_i - E_0 - \hbar\omega} + \frac{1}{E_i - E_0 + \hbar\omega} \right) . \tag{3.23}$$

The response of the system to the perturbation is seen to diverge when the perturbing frequency matches a transition energy to an excited state. Algebraically, the above-defined response function behaves in the same way as the classical response of a mass on a spring to an external force, eq. (2.34).

It is interesting to examine the high frequency limit of the response, an opposite limit to the static polarizability. Let us expand eq. (3.23) in inverse powers of ω:

$$\alpha_V(\omega) = \frac{2\sum <i|V|0>^2 (E_i - E_0)}{\omega^2}$$

$$+ \frac{2\sum <i|V|0>^2 (E_i - E_0)^3}{\omega^4} + ... \tag{3.24}$$

The numerator of the first term will be recognized as the energy-weighted sum rule. It is not surprising that it should play a role in the response in the high frequency limit, since we know that it describes the short-time behavior of the system. The numerator in the next term is like the energy-weighted sum rule, but is weighted with the cube instead of the first power of the energy. Let us formally define this sum with the commutator expectation value,

$$S_3 = 2 \sum_i <0|V|i>^2 (E_i - E_0)^3 \; = <0|[H,V][H,[H,V]]|0>.$$

We can only call this a sum rule if the right hand side is easy to evaluate. This is not necessarily the case, and in fact the sum can even be infinite. But there is one situation where S_3 is useful. That is when the Hamiltonian contains only very smooth potential fields, as happens for the Hartree–Fock and mean field Hamiltonians used in many calculations.

The major application of the S_3 sum is another formula for collective oscillation frequencies. Let us assume again that the field V excites only one state. Then its frequency may be calculated from the ratio of S_3 to the ordinary energy-weighted sum rule,

$$\omega_{dia}^2 = \frac{<0|[H,V][H,[H,V]]|0>}{<0|V[H,V]|0>} . \tag{3.25}$$

This is called the diabatic estimate, denoted with the subscript *dia*, because it is the oscillation frequency associated with the high-frequency limit of the response. To derive eq. (3.25), we assume the response is dominated by a single transition, having the form

$$\alpha(\omega) = \frac{A}{\omega_{dia}^2 - \omega^2} .$$

Expanding this in inverse powers of ω and comparing with eq. (3.24), we obtain eq. (3.25).

The formula for the diabatic frequency is very similar to the Rayleigh variational principle. To see the connection, we define the displacement field $\vec{u}_0 = \nabla V/m$. Then the denominator is $m \int \rho_0 |u_0|^2 d^3r$ which is half the inertia. The numerator is much more complicated. The inner commutators are $[H,V] = u_0 \cdot \nabla$, so the numerator is second order in u_0, as is the potential energy function in eq. (2.18). However, there are many terms to consider in general, so it is not clear *a priori* which classical potential energy function corresponds to the quantum physics. In the remaining chapters, we will analyze specific cases.

We conclude this section with the simple example of the dipole motion that was discussed in Chap. 2. Let us assume we have two kinds of particles, n and p, and take the field as $V = \sum_n z_n - \sum_p z_p$. The first commutator only has contribution from the kinetic energy operator and is

$$[H, V] = \frac{\hbar}{m}\left(\sum_p \nabla_z^p - \sum_n \nabla_z^n\right) .$$

This is essentially the difference of the two current operators. We next need the commutator of this operator with the Hamiltonian. The gradient commutes with the kinetic energy operator, so the only nonvanishing contribution comes from the gradient of the potential. In mean field theory, the potential is the single-particle potential $V_{n,p}(r)$ and the commutator is $[H, [H, V]] = \pm(1/m)dV_{n,p}/dz$. The remaining gradient can be put onto the single-particle wave function, which becomes a gradient of the density when multiplied by the conjugate wave function and summed over particles. We can write the result as

$$\omega_{dia}^2 = \frac{1}{Am}\int d^3r \frac{d\rho_n}{dz}\frac{dV_p}{dz}$$

Here V_p is the potential acting on the particles p, due to particles of type n. Finally, if we assume that the single-particle potential depends only on the local density, we have

$$\frac{dV}{dz} = \frac{dV}{d\rho}\frac{d\rho}{dz} .$$

This is exactly equivalent to eq. (2.23), when we note that the single-particle potential may be expressed in terms of the potential energy density as $V_{p,n} = \partial v_{sym}/\partial\rho$.

4
RPA

In the last chapter we discussed in general terms the quantum mechanical description of collective oscillations of many-particle systems, without being very specific on the details of the wave functions. Here we will develop the theory further using the mean field theory of the many-particle wave function. The resulting theory of small amplitude oscillations is known as RPA*.

Mean field theory is one of the most useful approximations in all of physics and chemistry. In it we replace the many-particle Schrödinger equation by the single-particle Schrödinger equation. What started out as a problem too complicated to solve for all but the smallest systems becomes quite manageable when one deals only with one particle at a time. The many-body effects come in via the single-particle potential which is generated from the particles themselves. The resulting self-consistent mean field theory has been enormously successful in many domains of quantum mechanics. Whenever a many-particle system exhibits single-particle behavior, which happens quite commonly, mean field theory is likely to be useful. In App. A we give a short summary of the mean field theory as practiced in condensed matter physics and in nuclear physics.

The most familiar application of mean field theory is to describe stationary states, but the extension to time-dependent states is straightforward and leads directly to RPA. There are many derivations of RPA; we shall approach it from the solution of the time-dependent Schrödinger equation in an external field. One

---

* The name Random Phase Approximation, RPA, arose from the work of Bohm and Pines (1953) who derived the theory by assuming certain amplitudes had a random phase and could therefore be neglected.

looks at how observables such as density depend on the perturbation; this dependence is described by a response function which is nothing more than a generalization of the polarization introduced in Chap. 3. The response fu·iction formulation is convenient for many problems in which the excitation of the system by external fields is the main object of study. A quite different approach to RPA starts from Fock space operators and develops equations of motion from commutators with the Hamiltonian. This form is most convenient when the structure of the wave function of the oscillation is the primary object of study or is needed for some other reason. For example, when the interactions are very complicated, one needs a more explicit representation of the wave function than can be easily obtained by the response method.

4.1 Linear response

We will derive RPA from the response of a system to a weak external field. Let

$$H_0 = -\frac{\hbar^2}{2m}\nabla^2 + V$$

be the unperturbed, time-independent mean-field Hamiltonian. The associated time-dependent Schrödinger equation is

$$i\hbar\frac{\partial}{\partial t}\phi_i(t) = H_0\phi_i(t) = e_i\phi_i(t)$$

where ϕ_i is a single-particle wave function and e_i is its eigenenergy. The time dependence of the eigenstates $\phi_i(t)$ is very simple, namely $\phi_i(t) = \phi_i\exp(-ie_it)$. Here $\phi_i \equiv \phi_i(0)$ can be chosen to be real. The additional weak external field is a perturbation in the single-particle Schrödinger equation, and the change in the wave functions can be calculated by perturbation theory. First let us consider an external field V_x that is independent of time. Then the perturbed wave function is, to first order,

$$\phi_i' = (\phi_i + \sum_j \frac{<j|V_x|i>}{e_i - e_j}\phi_j)e^{-ie_it/\hbar} \quad .$$

The j sum runs over all single-particle states[†].

[†] However, the occupied states in the j-sum may be dropped without affecting any observables. Transitions to occupied states are eliminated when the antisymmetrized many-particle wave function is constructed.

We need to generalize the perturbation theory to time-dependent external fields. Let us give the external field a sinusoidal time dependence, $V_x(r) \cos \omega t$. One can then take the perturbed wavefunction to have the form

$$\phi_i' = \left[\phi_i + \sum_j (x_{ij} e^{-i\omega t} + y_{ij} e^{i\omega t})\phi_j\right] e^{-ie_i t/\hbar} \ .$$

The coefficients x_{ij} and y_{ij} are obtained by substituting the wavefunction in the perturbed Schrödinger equation,

$$i\hbar \frac{\partial \phi_i'}{\partial t} = -\frac{\hbar^2}{2m} \nabla^2 \phi_i' + V\phi_i + V_x(t)\phi_i' \ .$$

If the perturbation is turned on slowly at $t = -\infty$, the perturbed wave functions are found to be (Pines and Nozieres (1966))

$$\phi_i'(t) =$$

$$\left[\phi_i + \frac{1}{2}\sum_j <j|V_x|i> \left(\frac{e^{i\omega t}}{e_i - e_j - \hbar\omega} + \frac{e^{-i\omega t}}{e_i - e_j + \hbar\omega}\right)\phi_j\right] e^{-ie_i t/\hbar} \ .$$

This wavefunction can also be written as

$$\phi_i'(t) = \left(\phi_i + \phi_i^R + \phi_i^I\right) e^{-ie_i t/\hbar}$$

where

$$\phi_i^R = \sum_j \frac{(e_i - e_j) <j|V_x|i>}{(e_i - e_j)^2 - \omega^2}\phi_j \cos \omega t$$

and

$$\phi_i^I = \sum_j \frac{\omega <j|V_x|i>}{(e_i - e_j)^2 - \omega^2}\phi_j \sin \omega t \ .$$

The time-dependent density and current may easily be found from the wave function. Simple expressions are derived by neglecting terms of order V_x^2 and making use of the fact that ϕ_i is real. The results are

$$n(r,t) = \sum_i^{occ} |\phi_i'(t)|^2 = \sum_i (\phi_i)^2 + 2\sum_i \phi_i\phi_i^R$$

$$= n_0(r) + \delta n(r) \cos \omega t$$

with
$$\delta n(r) =$$

$$\sum_{i}^{occ} \sum_{j} <j|V_x|i><i|n(r)|j> \left(\frac{1}{e_i - e_j - \hbar\omega} + \frac{1}{e_i - e_j + \hbar\omega} \right)$$

and

$$\vec{j}(r,t) = \frac{\hbar}{im} \sum_i \text{Im}(\phi_i \vec{\nabla} \phi_i^I - \phi_i^I \vec{\nabla} \phi_i)$$

$$= \sin \omega t$$

$$\times \sum_{i}^{occ} \sum_{j} <j|V_x|i><i|\vec{j}|j> \left(\frac{1}{e_i - e_j - \hbar\omega} - \frac{1}{e_i - e_j + \hbar\omega} \right).$$

The matrix elements of the density operator $<i|n(r)|j>$ and the current operator $<i|\vec{j}|j>$ were defined in eq. (3.1–2). Let us now define a response function for the density change at r induced by a potential field at r', leaving out the explicit time dependence:

$$\Pi^0(r,r',\omega) = \sum_{i}^{occ} \sum_{j} <i|n(r)|j><j|n(r')|i>$$

$$\times \left(\frac{1}{e_i - e_j - \hbar\omega} + \frac{1}{e_i - e_j + \hbar\omega} \right). \tag{4.1}$$

Then the response to an arbitrary external field can be calculated from the integral,

$$\delta n_{ip} = \int \Pi^0(r,r',\omega) V_x(r') d^3 r'.$$

This is the independent particle response, which we indicated with a subscript ip. One more thing has to be added to make the RPA theory of the response. So far the only perturbation in the single-particle Schrödinger equation was the external field. However, the induced density oscillation will cause the self-consistent field to oscillate at the same frequency as well. Given the dependence of the potential field on particle density, we can evaluate the change with respect to small density changes. Let us define the functional derivative $\delta V(r)/\delta n(r')$ which gives this dependence. The time-varying mean field is then

$$\delta V(r) = \int d^3 r' \frac{\delta V(r)}{\delta n(r')} \delta n(r') \quad .$$

Adding this potential to the external potential, we obtain an implicit equation for the self-consistent density,

$$\delta n_{RPA}(r) = \int \Pi^0(r, r_2)\left(V_x(r_2) + \int d^3r' \frac{\delta V(r_2)}{\delta n(r')} \delta n_{RPA}(r')\right) d^3r_2 \ .$$

We next define an RPA response function,

$$\delta n_{RPA} = \int \Pi^{RPA}(r, r', \omega) V_x(r') d^3r' \ .$$

Before the moment induced in the system by an external field was expressed in terms of the total field, but in the last equation we try to express it directly in terms of the external field. This can be achieved if Π^{RPA} satisfies the implicit equation,

$$\Pi^{RPA}(r, r', \omega) = \Pi^0(r, r', \omega)$$
$$+ \int d^3r_2 d^3r_3 \Pi^0(r, r_2, \omega) \frac{\delta V(r_2)}{\delta n(r_3)} \Pi^{RPA}(r_3, r', \omega). \quad (4.2)$$

This is an integral equation which we write more compactly as an operator equation (Fetter and Walecka (1971)),

$$\Pi^{RPA} = \Pi^0 + \Pi^0 \frac{\delta V}{\delta n} \Pi^{RPA} \ .$$

This operator equation has the formal solution,

$$\Pi^{RPA} = \left(1 - \Pi^0 \frac{\delta V}{\delta n}\right)^{-1} \Pi^0 \quad \text{(RPA response)}, \quad (4.3)$$

which is the basic equation for the RPA response. In practical terms, the operator equation is often replaced by a matrix equation, representing $\Pi(r, r')$ on a spatial mesh. Then the solution is obtained by inverting the matrix representing $(1 - \Pi^0 \delta V/\delta n)$. The eigenfrequencies of the system are determined by the poles of (4.2), while the transition matrix elements are related to the residues.

Let us define the states of the system associated with the RPA excitations, $|a>$. Then analogous to eq. (4.1) one can express the RPA response in the form

$$\Pi^{RPA}(\vec{r}, \vec{r}', \omega) = \sum_a <0|n(r)|a><a|n(r')|0>$$

$$\times \left(\frac{1}{E_a - E_0 - \hbar\omega} + \frac{1}{E_a - E_0 + \hbar\omega}\right)$$

where $|0>$ is the corresponding ground state. The associated energies are $E_a = \hbar\omega_a$.

An important property of the response is preserved by the RPA. If we take the limit of the response for $\omega \to \infty$, then $(1 - \Pi^0 \delta V / \delta n)^{-1} \to 1$ for ordinary potentials. This implies from eq. (4.2),

$$\lim_{\omega \to \infty} \Pi^{RPA} \to \Pi_0(r, r', \omega) \ .$$

Now expand Π_0 and Π^{RPA} in powers of $1/\omega$, and use the above limit to equate the leading terms. This yields two sums over transition densities which can be brought to a more familiar form by integrating out the coordinates with the fields $V_x(r)$ and $V_x(r')$,

$$\sum_{ij} (e_i - e_j) <i|V_x|j>^2 = \sum_a (E_a - E_0) <a|V_x|0>^2 \ .$$

That is, the energy-weighted sum does not depend on the residual interactions of the system. This is to be expected of a good theory, in view of the sum rule constraints(cf. Sec. 3.2).

Strength function

The excitation of systems by a short-duration perturbation is conveniently described in terms of the **strength function** S, defined by

$$S(V_x, \omega) = \sum_f <f|V_x|0>^2 \delta(\hbar\omega - E_f + E_0) \ . \qquad (4.4)$$

Here V_x is the spatial function describing the perturbation, f labels eigenstates of the system, and $E_f - E_0$ is the associated excitation energy. It is interesting to examine the situation where there are a limited number of collective states embedded in a large set of more complicated states. The full solution of the Hamiltonian problem gives the eigenstates $|f>$. These contain as components the collective states $|a>$. If there is only one collective state, the strength function can be written

$$S(V_x, \omega) = <a|V_x|0>^2 P_a(\omega)$$

where we have defined a strength function for the state $|a>$,

$$P_a(\omega) = \sum_f (<f|a>)^2 \delta(\hbar\omega - E_f + E_0) \ .$$

The quantity P_a gives the probability of finding the state per unit energy interval of the spectrum.

The strength function is simply related to the response, proportional to its imaginary part:

$$S(V_x, \omega) = \frac{1}{\pi} \int d^3 r d^3 r' V_x(r) V_x(r') \mathrm{Im} \Pi(r, r', \omega) \ . \tag{4.5}$$

In this context, the function $\Pi(r, r', \omega)$ is given an imaginary part either by adding a small term $i\eta$ to ω, or, in the case of continuum states, using outgoing wave boundary conditions in the definition of the single-particle wave functions ϕ_j. The above relation is easy to derive for the unperturbed response Π^0, using the relation $\mathrm{Im}(1/(x - i\eta)) = \pi\delta(x)$. Eq. (4.4) is the quantum mechanical analog to eq. (2.35), giving the power transfer rate to a classical system. In fact, according to Fermi's golden rule of second-order perturbation theory, the probability per unit time that the internal field V_x transfers energy $\hbar\omega$ to the system is given by

$$P(V_x, \omega) = \frac{2\pi}{\hbar} \sum_f <f|V_x|0>^2 \delta(\hbar\omega - E_f + E_0)$$

$$= \frac{2\pi}{\hbar} S(V_x, \omega) \ .$$

Consequently, the energy transfer per unit time is equal to

$$\frac{dE}{dt} = 2\pi\omega S(V_x, \omega) \ .$$

4.2 Matrix formulation of RPA

The above response formalism gives no possibility to treat the exchange interaction properly, because it depends on having a local interaction in the single-particle Schrödinger equation. A more general formulation may be made using a configuration space representation of the operators. Here we specify the perturbation by the initial and final orbitals rather than by the density change. The independent particle response is now defined

$$\Pi^0(ij, kl, \omega) = \delta_{ik}\delta_{jl} \frac{n_i - n_j}{e_i - e_j - \hbar\omega} \tag{4.6}$$

where $n_i = 0$ or 1 depending on whether the state i is empty or occupied in the ground state. We have written the response explicitly as a matrix in the space of orbital pairs (i, j). Note that it is nonzero only if one of the partners of the pair (i, j) is a particle and the other is a hole. The reader may want to check that this definition is consistent with our previous definition of the independent particle response in coordinate space, eq. (4.1). In fact, the coordinate space response can be derived from the configuration space response simply by expanding the equation

$$\Pi^0(r,r',\omega) = \sum_{ij,kl} <i|n(r)|j> \Pi^0(ij,kl,\omega) <k|n(r')|l> \quad .$$

The RPA response in the configuration representation may be derived from a similar equation relating the two representations,

$$\Pi^{RPA}(r,r',\omega) = \sum_{ij,kl} <i|n(r)|j> \Pi^{RPA}(ij,kl) <k|n(r')|l> \quad .$$

Inserting this in eq. (4.2) one can show that the RPA equation in the configuration representation has the same form as eq. (4.3), but with the residual interaction in that equation replaced by the interaction matrix in the configuration representation. This has matrix elements

$$<ij|v|kl> = \int d^3r d^3r' \phi_i^*(r)\phi_j(r)\frac{\delta V(r)}{\delta n(r')}\phi_l(r')\phi_k^*(r') \quad .$$

It is often convenient to express the RPA response in a diagonal form in the configuration representation. This will lead us to matrix diagonalization equations for RPA, which may seem at first somewhat mysterious because they contain negative signs not present in the diagonalization of ordinary Hermitean Hamiltonians. Our derivation will start from the singularities of Π^{RPA} as a function of ω, which of course will be present in any representation. The fact that there is a pole in the response at some frequency ω_a implies that there is a vector $x_a(ij)$ which satisfies the matrix equation

$$\left(1 - \Pi^0\frac{\delta V(r)}{\delta n(r')}\right)x_a = 0.$$

Looking at the (i,j) row of this equation, we may see with the help of eq. (4.6) that it may be written

$$(e_i - e_j - \hbar\omega_a)x_a(ij) + (n_i - n_j)\sum_{kl} <ij|v|kl> x_a(kl) = 0$$

$$\text{(RPA matrix).} \tag{4.7}$$

This looks exactly like an ordinary eigenvalue equation except for the factor $n_i - n_j$, which can be ± 1, depending on whether the orbitals (i, j) have particle-hole or hole-particle character. In this representation the response function may be written

$$\Pi^{RPA}(ij, kl) = \sum_a \frac{2\omega_a x_a(ij) x_a(kl)}{\omega^2 - \omega_a^2}$$

where now the normalization of the eigenvectors must be specified. We will not derive this, but the required normalization is

$$\sum_{ij} (n_i - n_j) x_a^2(ij) = 1.$$

It is common to use a notation with particle-hole amplitudes $(n_i - n_j = +1)$ denoted by $X_a(ij) = x_a(ij)$ and hole-particle amplitudes $(n_i - n_j = -1)$ by $Y_a(ij) = x_a(ji)$. The matrix eigenvalue equation then has the following block form,

$$\begin{pmatrix} A & B \\ -B & -A \end{pmatrix} \begin{pmatrix} X \\ Y \end{pmatrix} = \hbar\omega \begin{pmatrix} X \\ Y \end{pmatrix} \tag{4.8}$$

with

$$A_{ph,p'h'} = (e_p - e_h)\delta_{p,p'}\delta_{h,h'} + <ph|v|p'h'>$$

and

$$B_{ph,p'h'} = <hp|v|p'h'> \quad .$$

The corresponding normalization equation is

$$\sum_{ph} X_a(ph)^2 - Y_a(ph)^2 = 1 \tag{4.9}$$

In terms of the eigenvector amplitudes, the transition density for the vibration a is given by

$$\delta n(r) = \sum_{ij} <i|n(r)|j> x_a(ij)$$

or

$$\delta n(r) = \sum_{ph} \left(<p|n(r)|h> X_{ph} + <h|n(r)|p> Y_{ph} \right) \quad . \tag{4.10}$$

The transition potential of the vibration may be defined similarly, making use of eq. (2.17). Its matrix elements are

$$<p|\delta V|h> =$$

$$\sum_{p'h'} \left(X_a(p'h') <ph|v|p'h'> + Y_a(p'h') <ph|v|h'p'> \right)$$

Given the transition potential of the vibration, the X and Y amplitudes may be recovered from the eigenvalue equation (4.7), giving

$$X_a(ph) = -\frac{<p|\delta V|h>}{e_p - e_h - \hbar\omega_\alpha}$$

$$Y_a(ph) = -\frac{<h|\delta V|p>}{e_p - e_h + \hbar\omega_\alpha} \ . \tag{4.11}$$

It is useful to illustrate the above with a very simple example, namely the case when there is only one particle-hole configuration and $< ph|v|ph >=< ph|v|hp >$. Then V, X and Y are numbers rather than matrices or vectors, and the eigenvalue equation reads

$$eX + V(X + Y) = \hbar\omega X$$

$$-eY - V(X + Y) = \hbar\omega Y \ .$$

Here we have abbreviated $e_p - e_h = e$. For some purposes it is useful to rearrange the above equations by adding and subtracting them. This gives

$$(e + 2V)(X + Y) = \hbar\omega(X - Y)$$

$$e(X - Y) = \hbar\omega(X + Y) \ .$$

The $X + Y$ is proportional to a density and the $X - Y$ is proportional to a current. The first of these equations behaves like Newton's equation, relating the time derivative of the current to forces which depend linearly on a density perturbation. The second equation is a consequence of the continuity condition which relates the time rate of change of density to the current. The two equations can only be satisfied simultaneously if the frequency ω obeys the relation,

$$\hbar^2\omega^2 = e^2 + 2Ve \ . \tag{4.12}$$

Several obvious points may be made with this equation. First of all, the sign of V determines the direction of the energy shift with respect to the particle-hole state; an attractive V lowers ω while a repulsive interaction raises it. Second, if V is weak, the change in ω is small and is just the expected perturbation, $\hbar\omega \approx e + V$.

We may determine the X and Y amplitudes using eq. (4.9) and (4.11). The result is

$$X = -\sqrt{\frac{e}{\omega}} \frac{V}{e - \hbar\omega}$$

$$Y = -\sqrt{\frac{e}{\omega}} \frac{V}{e + \hbar\omega} .$$

The transition strength for an operator may now be determined. Suppose we have an operator M with the particle-hole matrix element $m = <p|M|h> = <h|M|p>$. Then from eq. (4.11) we find the collective state transition strength

$$<0|M|a>^2 = m^2(X + Y)^2 = m^2 \frac{e}{\hbar\omega} .$$

Thus excitations which are shifted to lower energy by the interaction have a larger transition strength. On the other hand, the oscillator strength associated with the transition is proportional to the product $<0|M|a>^2 \omega = m^2 e$, and is thus independent of the residual interaction.

4.3   Sum rules

An important characteristic of RPA is that it automatically satisfies the energy-weighted sum rules. It is easy to see why this is so in theories that have simple local potentials. The independent particle response, Π^0, represents nothing more than single-particle motion in an external potential. The sum rule for the field $F(r)$ may be expressed in terms of the strength function,

$$\int d\omega\, \omega S(F,\omega) = \frac{1}{\pi} \int d\omega\, d^3r d^3r' \omega F(r) \mathrm{Im}\Pi^0(r,r',\omega) F(r')$$

$$= \frac{1}{2m} \int d^3r n_0(r) |\nabla F|^2.$$

This is easy to prove using eq. (4.4) to write the integral as a sum over single-particle matrix elements. The factor ω times the single-particle matrix element is expressed as the commutator of the single-particle Hamiltonian with the field F, and finally the sum over final states is replaced by unity.

The corresponding sum rule for RPA simply has Π^0 replaced by Π^{RPA}. It has the same value as the independent particle sum rule when the interaction has a normal local form, as we saw earlier. This result is most obvious from the point of view of the time-dependent Schrödinger equation. The sum rule measures the energy absorbed immediately after an impulsive excitation. This energy is conserved in the time-dependent mean field approximation, so the sum rule has the same value in the independent particle model as in the self-consistent mean field theory. Since RPA is just the small amplitude limit of the time-dependent mean field theory, it must also preserve the sum rules.

The situation becomes more complicated if the interactions depend on momentum. Then the commutator of the Hamiltonian with the external field has more terms than just the one coming from the free kinetic energy operator. Thus in general, neither the free strength function nor the RPA strength function will satisfy the sum rules of Chap. 3. However, there is one case where these sum rules still apply. If the external field acts equally on all particles in the system, the double commutator of the field with the two-particle interaction vanishes. In this case, the sum rule would not be satisfied in the independent particle strength function but would be restored in the RPA strength function.

4.4   Separable interactions

We see that in general the RPA theory requires solving an operator equation. In the configuration representation, the corresponding matrix equation has the dimension of the number of particle-hole states, which can be enormous. Even in the coordinate space representation, the matrices are not small and only computer solutions are possible. However, if the interaction is separable the equation can be reduced to a purely algebraic form which can be solved analytically. By separable interaction we mean one that can be written as a product of two factors depending separately on r and r', i.e.

$$v(r, r') = \kappa F(r) F(r') \ . \tag{4.13}$$

Inserting this into the RPA response function, we obtain immediately

$$\Pi^{RPA}(r, r', \omega) =$$

$$\int d^3 r'' \left(1 - \kappa F(r_1) \int d^3 r_2 f(r_2) \Pi^0(r_2, r_3, \omega)\right)^{-1}_{r, r''} \Pi^0(r'', r', \omega) .$$

The operator inversion is carried out using a simple matrix identity for a matrix of the form $(1 + vw^\dagger)$, where v and w are vectors and $vw^\dagger$ is the dyadic (matrix) product of the vectors. The inverse of this matrix is given by

$$(1 + vw^\dagger)^{-1} = 1 + v \frac{1}{1 + v \cdot w} w^\dagger .$$

This may be easily verified multiplying the matrix and its inverse to obtain the unit matrix. For the RPA matrix inversion, we take $v = \kappa F$ and $w = \int d^3 r' F(r') \Pi^0(r, r', \omega)$. The resonances of the RPA theory are found at the singularities of the above function, which occur when the denominator $1 + v \cdot w$ vanishes. Thus the condition for a resonance is

$$0 = 1 + \kappa \int d^3 r d^3 r' F(r) \Pi^0(r', r, \omega) F(r') . \qquad (4.14)$$

This is called the RPA **dispersion relation**. Note that the integral in this formula is nothing more than the independent-particle approximation to the polarizability in the field F. Let us call the integral Π^0_F, that is

$$\Pi^0_F = \int d^3 r d^3 r' F(r) F(r') \Pi^0(r, r', \omega)$$

$$= \sum_{ph} <p|F|h>^2 \left(\frac{1}{e_p - e_h - \hbar\omega} + \frac{1}{e_p - e_h + \hbar\omega}\right).$$

The dispersion relation then reads

$$\Pi^0_F(\omega) = -\frac{1}{\kappa} . \qquad (4.15)$$

The qualitative behavior of the RPA resonances may be seen from a graphical solution of the dispersion relation (4.14). The left hand side of eq. (4.15) is indicated in Fig. 4.1 as the solid line on the left hand side. The function has poles at energies corresponding to particle-hole excitations, $e_p - e_h$. The solutions to the equation are the frequencies where the function equals $-\kappa^{-1}$. This is quite different depending on whether the interaction is attractive or repulsive. For a repulsive interaction, there is a solution at an energy higher than all the particle-hole energy differences, as depicted by the circled dot in the lower right hand graph in Fig. 4.1. This is the collective vibration; it will have a large transition strength for the field F. There are also other solutions to the dispersion relation located near the unperturbed particle-hole energies, indicated with solid dots. A simple formula can be given for the frequency of the collective state in the limit where all the particle-hole states are degenerate in energy. Denoting that energy by e, eq. (4.15) may be solved to obtain

$$\hbar^2 \omega_c^2 = e^2 + 2e\kappa \sum_{ph} <p|F|h>^2 \quad . \tag{4.16}$$

Note the close resemblance to eq. (4.12).

When the interaction is attractive, there will be a resonance lower in energy than the lowest particle-hole state, provided the interaction is not too strong. This is indicated in Fig. 4.1 on the upper right graph. Again, there are also other roots among the unperturbed particle-hole poles. As the interaction strength is increased, the low-frequency resonance goes to zero frequency, and for stronger interactions there is no real solution. This regime corresponds to a phase transition in mean field theory. For such strong interaction strengths, the ground state of the mean field Hamiltonian has a nonvanishing expectation value for the operator F.

Transition strengths

The transition strengths for individual resonances R associated with a field F may be calculated from the polarization propagator using eqs. (4.4,4.5),

$$<0|F|a>^2 = \mathrm{Im}\frac{1}{\pi} \int \Pi_F^{RPA} d\omega \quad .$$

Here the integral is over a small region near the pole at ω_a, and consequently receives the contribution of the residue. The residue of the pole in Π_F^{RPA} has a simple expression for separable interactions,

$$\left(\frac{d}{d\omega}\frac{1}{\Pi_F^{RPA}}\right)^{-1} = \left(\kappa^2 \frac{d}{d\omega}\Pi_F^0\right)^{-1} \quad . \tag{4.17}$$

Thus the transition strength is inversely proportional to the derivative of the independent particle polarization propagator. Look-

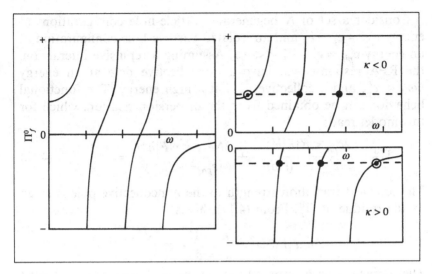

Fig. 4.1. Graphical solution of the RPA dispersion relation, eq. (4.14).

ing at Fig. 4.1, we see that Π^0 always has an especially small slope at an isolated solution of the dispersion relation, and therefore these solutions correspond to collective excitations. The collectivity may also be seen by explicitly constructing the transition matrix element from the X and Y amplitudes of eq. (4.11). All the configurations add with the same sign for the collective state, but there is destructive interference at the other solutions.

For repulsive interactions, the strength of the non-collective solutions with frequencies below that of the resonance is diminished below the independent-particle strength by the interaction, since it has shifted transition strength into the collective resonance. On the other hand, non-collective solutions lying higher than the resonance display an enhanced strength as compared to the single-particle value. This may be described as a frequency-dependent screening of the fields acting on the single-particle states by the collective resonance.

Screening

If we know the properties of the collective state, it is straightforward to estimate its screening effect on noncollective transitions In the following we use a schematic model to work out a simple estimate of the effect, given by eq. (4.20) below.

Consider a set of N degenerate particle-hole configurations at energy $e_{ph} = e_p - e_h$ and an isolated particle-hole configuration at an energy $e'_{ph} = e'_p - e'_h \ll e_{ph}$. Assuming a repulsive interaction, the RPA response will have a noncollective pole at an energy $\hbar\omega_{a'} \approx e'_{ph}$ and a collective pole at a high energy. The functional behavior can be obtained from the dispersion relation, which for this model reads

$$\Pi_F^0 = \frac{2e'_{ph} <p'|F|h'>^2}{(e'_{ph})^2 - (\hbar\omega)^2} + \frac{2Ne_{ph} <p|F|h>^2}{(e_{ph})^2 - (\hbar\omega)^2} = -\frac{1}{\kappa} \qquad (4.18)$$

The screened transition strength to the noncollective pole is given by the residue of Π_F^0. From (4.17) this is

$$<a'|F|0>^2 = \frac{\hbar}{\kappa^2} \left(\frac{\partial\Pi_F^0}{\partial\omega}(\omega'_a) \right)^{-1}$$

The derivative with respect to ω can be approximately evaluated as

$$\frac{\partial\Pi_F^0}{\partial\omega}(\omega'_a) \approx \frac{4e'_{ph}\hbar\omega'_a <p'|F|h'>^2}{((e'_{ph})^2 - (\hbar\omega'_a)^2)^2} \qquad (4.19)$$

as the contribution arising from the second term in eq. (4.18) is much smaller than that of the first when $\hbar\omega'_a \approx e'_{ph}$. Making use of the dispersion relation (4.18) one can rewrite the right hand side term of the above expression as

$$\frac{\hbar\omega_{a'}}{e'_{ph}} \frac{1}{<p'|F|h'>^2} \times \left(\frac{2Ne_{ph} <p|F|h>^2}{e_{ph}^2 - (\hbar\omega'_a)^2} + \frac{1}{\kappa} \right)^2 .$$

To get a final formula, we first note that $\hbar\omega'_a/e'_{ph} \approx 1$. We also rewrite the term in parentheses in the above expression using eq. (4.18) with $\omega = \omega_c$. At this frequency one can neglect the small contribution from the $(p'h')$ transition and write

$$2Ne_{ph} <p|F|h>^2 \approx -\frac{e_{ph} - (\hbar\omega_c)^2}{\kappa} .$$

When all the substitutions are made, one finally obtains

$$<a'|F|0>^2 \approx \frac{e_{ph}^2 - (\hbar\omega'_a)^2}{(\hbar\omega_c)^2} <p'|F|h'>^2 . \qquad (4.20)$$

This formula concisely displays the screening behavior: for $e'_{ph} \ll e_{ph} \ll \omega_c$ the transition strength is reduced by the factor $(e_{ph}/\omega_c)^2$.

Polarizability and collective potential

We return here to the static polarizability of the system, discussed in Sec. 3.6. In RPA, it is nothing more than the RPA polarization propagator evaluated at zero frequency,

$$\alpha_V(0) = \int d^3r\, d^3r'\, V(r)\Pi(r,r',0)V(r') \ .$$

Let us try to reduce this to classical physics, defining a collective coordinate q, and a force G associated with that coordinate. Assuming that the equilibrium is at $q = 0$, the lowest order expansion of the classical energy function is

$$E = \frac{1}{2}Kq^2 + Gq \ .$$

Minimizing this to find the polarization, we find

$$q = -\frac{1}{K}G \ .$$

This coefficient of proportionality between q and G is the static polarizability, which we will identify with $\alpha_V(0)$. In order to do this, we have to define q as the expectation value of the field V, and we have to define G as the coefficient of V in an external field, $V_{ex} = GV(r)$. Then the classical stiffness associated with this definition of coordinate and force is simply

$$K = \frac{1}{\alpha_V(0)} = \Pi_V(0)^{-1} \ .$$

When the force acting between particles is separable and we replace the field V by the separable force field F, the RPA expression becomes algebraic:

$$K = \left(\Pi_F^{RPA}(0)\right)^{-1} = \frac{1 - 2\kappa\sum_{ph} <p|F|h>^2 /(e_p - e_h)}{2\sum_{ph} <p|F|h>^2 /(e_p - e_h)} \qquad (4.21)$$

This expression will be used in Chap. 6.

The interaction

We have sketched the structure of the theory with separable interactions but not yet said how they may be introduced in a physical situation. There are two questions here: what is the appropriate form factor F for a given collective mode, and how is κ determined? Concerning the form factor F, we are strongly guided by nature, which tells us that many-particle systems commonly oscillate collectively with simple functions describing the displacement fields and the transition densities. Thus, given the kind of motion under study, F may be determined as the single-particle potential field that induces that motion. Given the form of F, there are a number of strategies to fix κ. We shall use self-consistency arguments when we consider surface vibrations. We can also start from the underlying two-particle interaction, and simply expand it in a complete set of single-particle fields including F as one of them. Ignoring the other terms besides $F(r)F(r')$ gives us the separable interaction. It may seem surprising, but in the RPA most of the interaction between particles is irrelevant to the motion, except for that part which produces an internal field having the same symmetry as the external field.

5
Dipole oscillations

The displacement of charge in a fixed direction produces a dipole moment in the system which couples strongly to electromagnetic fields. The displacement vector can be as simple as the uniform displacement field $\hat{z}$, making dipole modes relatively simple to construct and rather collective in the systems we find in nature. This collectivity, or concentration of a large fraction of the oscillator strength in a small frequency interval, gave this mode the name "giant dipole resonance" in nuclear physics, a term that is now applied to a corresponding bump that appears in the optical absorption spectrum of atoms as well. In this chapter we shall discuss classical and quantum mechanical models for the dipole vibrations. Our emphasis is on the forces that determine the average frequency. Other effects, such as damping of the vibrations, we defer to Chaps. 9 and 10. We begin with clusters of metal atoms, which show a collective dipole resonance known as the Mie resonance.

5.1 Dipole oscillations of electrons and the Mie theory

The free electron picture of metals leads to a similar model for the electronic structure of metallic clusters. The electrons are confined by a smooth potential that follows the spatial geometry of the cluster, but doesn't have any of the details of the interactions with individual atomic cores. The ground state in this model would be constructed by finding the single-electron eigenstates of the potential, and making a Slater determinant with the electrons in the lowest single-particle states. The dipole excitation strength

function would next be constructed in terms of transitions between the ground state and excited states of one-particle one-hole
character. However, this model would not describe the physical
situation, because the electron–electron interaction is left out. The
simplest quantum theory that corresponds to the actual behavior
is the RPA, and in fact the RPA also goes over into the classical
Mie theory in the limit where the electron density satisfies the
assumptions in the classical derivation.

The classical Mie theory produces a simple formula for the
frequency of the dipole oscillation, given by eq. (5.6) below. We
present a simple derivation using the Rayleigh variational principle, eq. (2.12). The starting assumption, which will be justified
a posteriori, is that the electron motion is just a uniform displacement. We take the displacement field to be in the z-direction, with
a magnitude a,

$$\vec{u}(r) = a\hat{z} \ .$$

The transition density is then given by eq. (3.3) as

$$\delta\rho(r) = a\nabla_z\rho_0 \ . \tag{5.1}$$

Let us now assume that the electron density is uniform inside a
sphere of radius R, and vanishes outside, $\rho_0(r) = n_0\Theta(R-r)$, where
$\Theta(x)$ is the step function equal to 1 for $x > 0$ and to 0 for $x < 0$.
The transition density then has the form

$$\delta\rho = a\,n_0\nabla_z\theta(R-r) = -a\,n_0\,P_1(\cos\theta)\delta(r-R) \tag{5.2}$$

where $P_1(\cos\theta) = \cos\theta$ is the Legendre polynomial of order 1. The
oscillating potential field produced by this distribution is found
by integrating the charge with the Coulomb interaction in the
multipole representation,

$$v_{coul}(r_1,r_2) = \frac{e^2}{|r_1 - r_2|} = \frac{e^2}{r_>}\sum_L \left(\frac{r_<}{r_>}\right)^L P_L(\cos(\theta_{12})) \ . \tag{5.3}$$

Physically, it is the field inside the cluster that will accelerate the
electrons while the fluctuating charge is located on the surface.
Thus, we apply the multipole decomposition taking the field coordinate as $r_<$ and the source coordinate as $r_> = R$. Because the
charge distribution is dipolar, only the $L = 1$ term in the sum
will contribute to the integral over charge. Thus the oscillating

single-particle field is (cf. eq. (2.17))

$$\delta V(\vec{r}) =$$

$$e^2 \sum_L \frac{r^L}{R^{L-1}} a n_0 \int P_1(\cos(\theta')) P_L(\cos(\theta - \theta')) d\Omega =$$

$$\frac{4\pi}{3} e^2 a r \cos(\theta) n_0 \quad (r < R).$$

The potential energy associated with this field and the transition charge density eq. (5.2) can be calculated using eq. (2.18). The integral is easily evaluated because the charge distribution is a delta function at the surface of the sphere*

$$C = \int \delta \rho \delta V d^3 r = \frac{2\pi}{3} a^2 n_0^2 e^2 R^3 \int \cos^2(\theta) d\Omega = \frac{2\pi}{3} a^2 e^2 n_0 N \quad .$$
(5.4)

In the last equation above, we used the relation between particle number N, density n_0, and volume of the sphere, $N = 4\pi n_0 R^3/3$.

We also need the inertia associated with this motion, which is given by

$$I = \int m n_0 \vec{u} \cdot \vec{u} d^3 r = N a^2$$
(5.5)

The Rayleigh formula eq. (2.12) then gives the frequency as the ratio of C to I,

$$\omega_M^2 = \frac{4\pi n_0 e^2}{3m} = \frac{e^2}{r_0^3 m} \quad .$$
(5.6)

This is the Mie formula for the dipole frequency, which we shall write as ω_M. The last expression relates ω_M to the Wigner–Seitz radius of the system, $r_0 = a_0 r_s$. The above expression can also be written in terms of the classical polarizability of a sphere, $\alpha = R^3$, as

$$\omega_m^2 = \frac{e^2}{m} \frac{N}{\alpha} \quad .$$
(5.7)

For sodium clusters $r_s \approx 4$ and $\omega_M = 3.4$ eV/$\hbar$. In Table 5.1 we display the Mie frequency for different metals, assuming only the valence electrons contribute. It is interesting to note that the collective behavior of the Mie resonance is present also in

---

* The potential is continuous at the surface, permitting the evaluation of the integral using the inner potential.

Table 5.1. Mie resonance energy in selected metallic elements, from eq. (5.6)

element	n (10^{22}/cm^3)	r_s	$\hbar\omega_M$ (eV)
Li	4.70	3.25	4.6
Na	2.65	3.93	3.5
K	1.40	4.86	2.5
Cs	0.91	5.62	2.0
Cu	8.47	2.67	6.2
Ag	5.86	3.02	5.2
Be	24.7	1.87	10.6
Mg	8.61	2.66	6.3
Ca	4.61	3.27	4.6
Ba	3.15	3.71	3.8
Fe	17.0	2.12	8.8
Zn	13.2	2.30	7.8
Cd	9.27	2.59	6.5
Hg	8.65	2.65	6.3
Al	18.1	2.07	9.1
Ga	15.4	2.19	8.4
In	11.5	2.41	7.3
Sn	14.8	2.22	8.2
Pb	13.2	2.30	7.8
Bi	14.1	2.25	8.1

nonmetallic materials. A timely example is the fullerine C_{60}, which might be thought of as a large, spherical graphite molecule. Detailed RPA calculations predict a collective dipole oscillation at 20 eV, which has been confirmed experimentally (Hertel et al. (1992)). A classical model for the Mie resonance in C_{60} could be made with the electron motion in a conducting spherical shell. This problem can be solved exactly (Lundqvist (1983)), and one finds that there are two modes, depending on whether the surface charges on the inner and outer layer oscillate with the same or the opposite phase. In fact there is also a lower mode on the C_{60} cluster with much smaller oscillator strength, the so-called π-electron plasmon.

Several features should be mentioned about the Mie formula. First, note that the frequency depends only on fundamental constants and the density of the electrons. The formula gives an unambiguous prediction; it must be satisfied in large spherical clusters if the following interdependent assumptions are valid:

- the nonvalence electrons are ignorable;

- the dipole strength is concentrated in a single mode of oscillation;

- the electron density is uniform in the interior.

In our derivation, we assumed that the displacement field was uniform, which in general would just be an approximation to the true motion. However, it turns out to be self-consistent in this case because the potential field arising from the displacement gives a uniform acceleration to the particles.

The oscillation in a bulk material, the plasmon, has an identical formula except for the factor of 3 in the denominator, that is

$$\omega_p^2 = \frac{4\pi n_0 e^2}{m} \ .$$

A general expression can be derived for displacement fields of arbitrary multipolarity L, which reads

$$\omega_{ML}^2 = \frac{L}{2L+1}\omega_p^2 \ .$$

In the limit where $L \to \infty$ the frequency approaches $\omega_p/\sqrt{2}$. This is the well-known surface plasmon of a conductor with a plane surface. We shall treat higher multipoles generally in Chap. 6.

We will also treat later in this chapter small ellipsoidal distortions of the sphere. The distortion raises or lowers the frequency depending on whether the oscillation is along a short or long axis.

Mie plasmon in RPA

The RPA theory provides a description of the vibration in terms of particle-hole excitations. If the interaction is treated as separable, the properties of the vibration can be calculated analytically. Let us assume that the transition density is surface-peaked, and that the particle energies are degenerate. With the first assumption one can write the component of the Coulomb interaction that couples to the dipolar motion along the z-axis in the form

$$v = \kappa D_1 D_2$$

where the coupling strength is given by $\kappa = e^2/R^3$ and the D are dipole fields, $D_i = z_i$. Making use of the dispersion relation eq. (4.16) and the TRK sum rule eq. (3.13) one can write

$$\omega^2 = \omega_0^2 + \frac{4\pi}{3} \frac{e^2 \hbar^2 n_0}{m}$$

where $\omega_0 = (e_p - e_h)/\hbar$ is the particle-hole energy. Neglecting ω_0^2 with respect to the second term, one obtains the Mie formula. From this result we see that the RPA theory adds minimal quantum effects to the classical theory. The most fundamental of these is the quantization of energy of the vibration in units of $\hbar\omega$. Then instead of discussing the time-varying density or potential in classical terms, one studies the transition density or transition potential between states with and without a quantum of excitation. The frequency will also differ from the Mie formula, and in fact the single oscillation can become fragmented into many frequencies. For the moment we will ignore the quantum effects on the frequency, and just examine the transition potential and densities.

If the dipole oscillation is indeed concentrated in a single state, the transition density will have the functional form given by eq. (5.1). Let us continue for the moment the approximation that the ground state density is sharp, giving eq. (5.2) for the transition density. The amplitude a can be determined from the TRK sum rule as follows. Using eq. (5.2) for the transition density, the dipole matrix element is

$$<i|z|0> = \int r \cos \theta d^3 r = aR^3 n_0 \int \cos^2 \theta d\Omega = Na \ .$$

Assuming now that only one state contributes to the sum rule eq. (3.13), it reads

$$\hbar\omega_M <M|z|0>^2 = \frac{\hbar^2}{2m} N.$$

Combining the last two equations and solving for a, we find

$$a = \sqrt{\frac{\hbar}{2m\omega_M N}} \ .$$

One could also derive the properties of the quantum transition density requantizing the harmonic oscillator that describes the classical motion. This yields the above relation for a as the zero-point amplitude of the vibrational motion.

We may evaluate the transition potential by convoluting the transition density with the Coulomb interaction,

$$\delta V = \int dr' \frac{e^2}{|r - r'|} \delta \rho(r') \ .$$

The details of this are exactly the same as in the last section. For completeness, we write here the potential both inside and outside the sphere.

$$\delta V(r) = e^2 n_0 \frac{4\pi}{3} a z \quad (r < R)$$

$$= e^2 n_0 \frac{4\pi}{3} a \frac{z}{r^2} \quad (r > R) \ .$$

Inserting the value of a, we obtain the formula for δV in terms of physical quantities:

$$\delta V(r) = z \sqrt{\frac{\omega_M^3 m\hbar}{2N}} \quad (r < R) \tag{5.8}$$

$$= \frac{z}{r^2} \sqrt{\frac{\omega_M^3 m\hbar}{2N}} \quad (r > R). \tag{5.9}$$

Beyond the Mie formula

We now consider quantum effects that will modify the frequency from the classical value. If the resonance is concentrated in a single state, we may use the diabatic formula, eq. (3.25), to calculate the frequency, which is very similar to the Rayleigh formula. One obvious effect will be the spill-out of electrons from the surface of the sphere, due to penetration of the electron wave function into the classically forbidden region, which will modify the integral, eq. (2.18). We may treat the integral more accurately using a parameterized function for the ground state electron distribution, and then using eq. (5.1) to determine $\delta \rho$. One possible parameterization of the ground state electron distribution is the Fermi function, $n_0(r) = n_0/(1 + \exp((r - R_0)/a_0))$. We may evaluate the integral eq. (2.18) as a power series in a_0. To first order the result is

$$C = \frac{4\pi N n_0 e^2}{3} \left(1 - \frac{a_0}{R_0} \log_e 2 \right)$$

Inserting this in the formula for the collective vibrational frequency gives

$$\omega^2 = \omega_M^2(1 - 0.69\frac{a_0}{R_0}).$$

Fitting a_0 to the density distribution of the jellium model, a typical value is $a_0 = 0.8$ au. For a sodium cluster with 8 atoms, the radius of the cluster is about $R_0 \approx 8$ au, so the spill-out produces a decrease of frequency by less than 4%. This is far too small to explain the experimental shift from the Mie frequency.

Another effect that both shifts the frequency of the resonance and affects the transition strength is the presence of core electrons in the system. The core electrons can be polarized slightly, decreasing the effective interaction strength between the valence electrons, and lowering the Mie resonance. This effect has been estimated by Pines (1963) for bulk plasmons, and it can be of the order of 5%. Note that the classical bulk plasmon formula often gives too high a frequency compared to measured bulk plasmons[†].

If the dipole excitation is split into several components, the diabatic formula gives a frequency whose square is the f-weighted average of the squared frequencies of the components. In practice, the multicomponent structure of the resonance makes the above calculation unreliable. One finds a strong component at low frequency and several small components at high frequency. The observed plasmon is usually identified with the low-frequency component.

In this situation, the polarizability sum provides a more useful estimate of the collective resonance frequency. The relation is identical to eq. (5.7), where now the polarizability α is the physical one. The relation is rather well satisfied in spherical sodium clusters, taking the frequency of the observed major peak. For example, Selby et al. (1989) found that the Mie resonance in Na_8 was centered at an energy of 2.53 eV. The polarizability of Na_8 has also been measured to have a value of 130 Å (Knight et al. (1985)). Eq. (5.7) then predicts a resonance value of 2.60 eV, which is higher than the observed by only 4%. This is in contrast to the Mie formula, whose prediction is 40% higher than observed.

[†] This leads back to Mie's original analysis, to express the frequency in terms of the bulk dielectric properties.

Deformation effects

As mentioned earlier, the plasmon frequency depends on the shape of the system. Ellipsoidal shape distortions are particularly effective in this regard, and their classical effect can be calculated analytically in terms of elliptic integrals. Since these distortions are important for considering nonspherical clusters and for modifying the plasmon in heated clusters, we show here how the effect can be calculated. However, we follow our original method and estimate the effects perturbatively rather than carry out the coordinate transformations necessary for the analytic calculation.

We begin with a ground state charge distribution that has a small quadrupole deformation. This may be expressed as

$$\rho_D = \rho_{sph} - \epsilon(3\cos^2\theta - 1)r\frac{d\rho_0}{dr}$$

where ρ_{sph} is the spherical density distribution and ϵ is the deformation defined in eq. (2.25) and in App. B. Then the collective dipole transition density is given by

$$\delta\rho_D = a\nabla_z\rho_D$$

$$= \delta\rho_{sph} + a\epsilon(5\cos\theta - 3\cos^3\theta)\frac{d\rho_0}{dr}$$

$$+a\epsilon(3\cos^3\theta - \cos\theta)r\frac{d^2\rho_0}{dr^2}$$

where $\delta\rho_{sph}$ is the transition density associated with the spherical system. We next find the transition potential produced by this charge distribution, using eq. (2.17). We only need the dipole component in the interior. It is given by

$$\delta V_D = e^2 r\cos\theta \int \frac{d^3r'}{(r')^2}\cos\theta'\delta\rho_D(\vec{r}')$$

$$= \delta V_{sph} + ae^2 r\cos\theta\epsilon\left[n_0 \int d\Omega'(5\cos^2\theta' - 3\cos^4\theta') + \int d\Omega'(\cos^2\theta' - 3\cos^4\theta')\int dr'r'\frac{d^2\rho_0}{dr^2}\right]$$

where δV_{sph} is the transition potential associated with the spherical system. The angular integrals are elementary and the last integral over the second derivative of the density is evaluated by parts. The final result for the dipolar component of the transition potential is

$$\delta V_D = \delta V_{sph}\left(1 - \frac{12\epsilon}{5}\right) .$$

The inertia in Rayleigh's principle only changes with ϵ quadratically, so the frequency dependence on ϵ is then

$$\omega = \omega_M\left(1 - \frac{6\epsilon}{5} + \mathcal{O}(\epsilon^2)\right) . \tag{5.10}$$

This is commonly expressed in terms of $\delta \approx 3\epsilon$, the relative difference between the largest and smallest axes of the ellipsoid. The above expression leads to

$$\omega = \omega_M\left(1 - \frac{2\delta}{5}\right)$$

to lowest order in the deformation parameter. The oscillations along the x- and y-axes are shifted upward in frequency by $\omega_M\delta/5$. Thus the total splitting between the lower component carrying 1/3 of transition strength and the upper component carrying 2/3 of the transition strength is $3\delta/5$.

In Fig. (5.1) we display the photoabsorption cross section associated with the cluster Na_{10}. According to the spheroidal cluster shell model this system is expected to have a prolate deformation, with deformation parameter $\delta \approx 0.4$, so the relative splitting between the resonances in predicted to be $3\delta/5 \approx 0.24$. From the figure, we see the two peaks separated by ≈ 110 nm with a centroid at ≈ 500 nm, with a ratio $110/500 \approx 0.22$ in good agreement with the simple Mie theory.

5.2   Nuclei

As we saw in Chap. 1, the giant dipole resonance is the most prominent feature of the photon absorption cross section in the energy range of 1–100 MeV. The systematics of the dipole frequency as a function of nuclear size are shown in Fig. 5.2. The resonance energy $\hbar\omega_D$ for heavy nuclei is given approximately by the empirical formula

$$\hbar\omega_D \sim \frac{80}{A^{1/3}} \quad \text{MeV.} \tag{5.11}$$

We know that the resonance has a dipolar character on several grounds. First, the wavelength of the photons is much larger

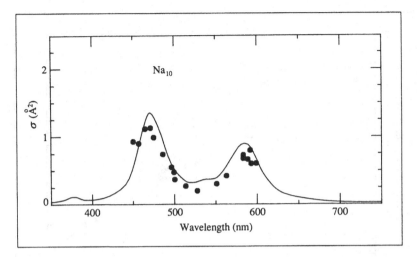

Fig. 5.1. Photoabsorption cross section in Na$_{10}$. The data are from Selby et al. (1989); the smooth curve is a theoretical fit by Bernath, Yannouleas and Broglia (1991).

than the size of the nucleus. This ratio may be estimated as $2\pi c/\omega(1.2 \text{ fm})A^{1/3} \sim 7$. Thus the long-wavelength approximation allowing only the dipole field should be reasonably satisfied. Second, the integrated cross section for this resonance comes close to half the oscillator strength of the protons. This is the theoretical maximum when the center-of-mass constraints are taken into account, as we will see in the next section. There are also more subtle ways to infer the character of the oscillation that we will not discuss, for example measuring the polarization of the emitted photons.

Center-of-mass effects

The fact that protons and neutrons have nearly the same mass makes a complication in the theory of nuclear dipole excitations. A uniform displacement field applied to all the particles of a system would simply move the center of mass without changing the internal structure at all. Any true excitation can be made without moving the center of mass. To see the consequences of this for the electric dipole operator and the oscillator strength, we write the operator as a sum of a term that acts only on the center

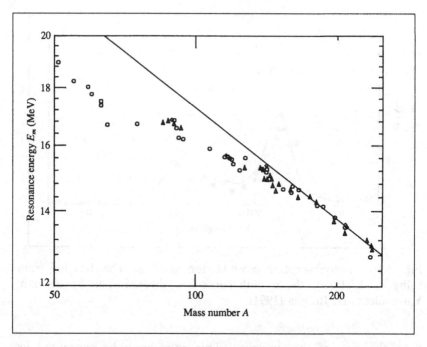

Fig. 5.2. Systematics of giant dipole energy, from Berman and Fultz (1975). The line shows the fit from eq. (5.11).

of mass and a remainder

$$e \sum_p z_p = e\left[\frac{N}{A}\sum_p z_p + \frac{Z}{A}\sum_p z_p + \frac{Z}{A}\sum_n z_n - \frac{Z}{A}\sum_n z_n\right]$$

$$= e\mathscr{E}\left[\frac{N}{A}\sum_p z_p - \frac{Z}{A}\sum_n z_n + Z\bar{z}\right]$$

where p stands for protons, n for neutrons, and

$$\bar{z} = \frac{\sum_i z_i}{A}$$

is the position of the center of mass of the nucleus. The term $e\mathscr{E}Z\bar{z}$ acts only on the center of mass of the nucleus and thus cannot excite it. It leads to nuclear Thomson scattering of photons. The other term,

$$D_{ex} = e\left[\frac{N}{A}\sum_p z_p - \frac{Z}{A}\sum_n z_n\right] \ , \qquad (5.12)$$

alone produces true excitations in the nucleus. Thus each proton acts as if it had an "effective charge" eN/A, and each neutron as if it had an "effective charge" $-eZ/A$.

The sum rule associated with the true excitations can be easily evaluated following the derivation for the ordinary dipole operator. The result is

$$\int \sigma dE = \frac{2\pi^2 \hbar^2 NZ}{Am},$$

which is half the usual sum rule when $N = Z$. As mentioned before, the remainder of the usual sum rule goes into an elastic process, nuclear Thomson scattering.

The dipole frequency

We described in Chap. 2 a model for the nuclear giant dipole very similar to the Mie theory. As in the Mie theory, the nuclear model assumed a uniform displacement field. However, the restoring force is quite different; it is now the short-ranged proton-attraction instead of the long-range Coulomb interaction. The predicted mass dependence of the giant dipole is $\hbar\omega_D \sim A^{-1/6}$, which is contrary to the empirical behavior in the heavier nuclei[‡]. The problem is that the nature of the restoring force favors a displacement field that is small or vanishes on the surface. Displacement of a neutron surface with respect to the protons has a large energy cost because the density change is very large. A more favored type of motion in large systems is a motion in which the proton density is increased on one side of the nucleus, decreased on the other side, without moving protons through the surface. This can be achieved with a displacement field that satisfies the hydrodynamic equations for a compressible medium, a model first proposed by Steinwedel and Jensen (1950). The dipolar hydrodynamic field can be expressed in terms of the spherical Bessel function $j_1(x)$ as

$$\vec{u}(r) = \nabla j_1(qr) Y_{1,m}(\hat{r}) \ .$$

Here the boundary condition that protons do not move through the surface is satisfied by requiring $u_r(R) = 0$. This occurs at the zero of the derivative of the j_1 Bessel function; the condition is $qR \approx 2.46$.

[‡] However, the A-dependence in lighter nuclei approaches this very weak power; see Fig. 5.2.

For a hydrodynamic mode, the frequency is proportional to the reduced wavenumber q, so the size dependence is predicted to be $\omega_D \sim 1/R \sim A^{-1/3}$, in agreement with the empirical behavior in heavy nuclei.

The main physics can be extracted from a much less detailed model based on the adiabatic frequency estimate, eq. (3.22), which was first done by Migdal (1944). The adiabatic formula requires the polarizability of the system, and the major task is to estimate this quantity. We shall assume that the nucleus behaves as a compressible liquid, and allow the protons and neutrons to be redistributed in a way to minimize the total energy. For simplicity, we consider equal numbers of neutrons and protons, $Z/A \sim N/A \sim 1/2$.

We examine the polarization of the density distribution, which is expressed in terms of the quantity ρ_{sym},

$$\rho_{sym} = \rho_p - \rho_n.$$

The polarization density will be determined by minimizing the sum of the Coulomb and the nuclear energy. The Coulomb energy is given by

$$E_{Coul} = \frac{1}{2} e\mathscr{E} \int z \rho_{sym} d^3 r.$$

The nuclear energy favors equal numbers of neutrons and protons. This appears in the liquid drop formula for nuclear binding as the term

$$E_{sym} = b_{sym} \frac{(N - Z)^2}{2A}$$

where b_{sym} has the empirical value of 50 MeV. This term would arise from an energy density function proportional to ρ_{sym}^2, for example $\frac{1}{2} b_{sym} \rho_{sym}^2 / \rho_0$. The total energy is then given by the integral

$$E = E_{coul} + E_{sym} = \int \left(\frac{b_{sym}}{2} \frac{\rho_{sym}^2}{\rho_0} + e\mathscr{E} \rho_{sym} \right) d^3 r.$$

The minimum of this expression occurs when the functional derivative with respect to ρ_{sym} vanishes. This requires

$$\rho_{sym} = \frac{\mathscr{E} e \rho_0}{2 b_{sym}} z \ .$$

We can now determine the polarizability as the ratio of the dipole moment to the electric field. This is

$$\alpha = \frac{(e/2) \int z \delta \rho d^3 r}{\mathscr{E}}$$

$$= \frac{e^2 \rho_0 <r^2>}{12 b_{sym}} \ .$$

The ratio of the TRK sum rule and the polarizability sum gives the estimate of the dipole frequency. Remembering to use the center-of-mass corrected sum rule, we find

$$(\hbar \omega_D)^2 = \frac{e^2 \hbar^2 NZ/2AM}{\alpha/2} = \frac{3 \hbar^2 b_{sym}}{M <r^2>} \ . \tag{5.13}$$

The numerical evaluation of this expression, using $b_{sym} = 50$ MeV, $<r^2> = 3/5 \ (1.2 A^{1/3} \ \text{fm})^2$, yields eq. (5.11) with a coefficient of 93 instead of 80 MeV. This is certainly a closer agreement than one should expect from such a simple model.

Eq. (5.13) was derived assuming the nucleus to be spherical, but it is easy to generalize the formula to deformed nuclei. The inertia is independent of the shape, but the polarizability is larger in the direction of the longer axis. For an oscillation in the z-direction, we should replace $3/ <r^2>$ in the Migdal formula $1/ <z^2>$. For a prolate nucleus with deformation ϵ (cf. eq. (2.25)) the dipole frequencies become

$$\omega_z = \omega_D \exp(-2\epsilon)$$

$$\omega_x = \omega_y = \omega_D \exp(\epsilon) \ . \tag{5.14}$$

Thus the fractional splitting between the two modes is $3\epsilon = \delta \approx 0.95\beta$, with the upper mode roughly twice the strength of the lower. Experimentally, one finds in rare earth nuclei large enough deformations to show this effect. The photoabsorption cross section associated with the nucleus ^{160}Gd is shown in Fig. 5.3. We see a split resonance with the lower resonance at 12 MeV and the upper one at 16 MeV. The centroid energy is about 14.5 MeV. The fractional splitting is thus $\delta \approx (16 - 12)/14.5 = 0.27$. This number compares reasonably well with the number $\delta = 0.35$ extracted from the low-energy electromagnetic properties of the nucleus, in particular the quadrupole transition matrix elements of the ground state rotational band.

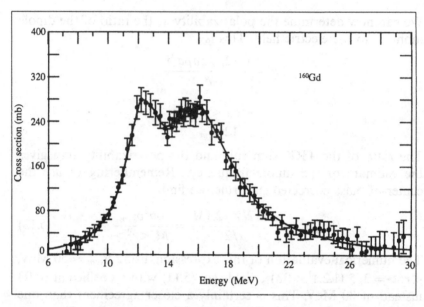

Fig. 5.3. Photoabsorption cross section in ^{160}Gd from Berman and Fultz (1975).

Microscopic estimate of dipole frequency

The estimate in the previous section treated the nucleus as a fluid drop without internal structure. Here we want to describe the vibration as a particle-hole excitation using RPA theory. In general, RPA requires computer resources to evaluate the particle wave functions and the interaction matrix elements. However, schematic models have been developed in which everything is calculated analytically, and we shall follow this approach. We first need to describe the single-particle wave functions and energies in a static nuclear potential. The harmonic oscillator provides a simple but useful approximation to this aspect of the calculation. From App. A, eq. (A.2), the nuclear size is fit with a an oscillator potential whose frequency is given by $\hbar\omega_0 \approx 41/A^{1/3}$ MeV. The lowest dipolar excitations are made by promoting a particle from a filled orbit in one shell into an empty orbit in the next higher shell. The excitation energy of the particle-hole state is just the oscillator energy $\hbar\omega_0$. This number is only half the giant dipole energy, so it is obvious that the interaction must play an important role. The interaction is only easy to calculate if we can make a separable approximation. This means that the interaction between two particles has the form

$$v(r_1, r_2) = \kappa F(r_1) F(r_2) \qquad (5.15)$$

where F is some function to be determined. It might seem that the above approximation is far too crude to be useful. But as mentioned in Chap. 4, most of the interaction between particles is irrelevant to the motion. The only important part is that which produces an internal field with the same symmetry as the external field exciting the motion. Thus, we could write any v as a sum of terms of the type shown above, and the only ones of importance would be the terms with dipolar symmetry in F.

To determine the optimum form for $F(r)$, we should know what the transition potential is in the giant dipole, and give F the same shape. Here we shall be rather crude and take the potential associated with a uniform displacement field,

$$F_{GD}(r) = z \tau_z \quad .$$

From the discussion in the previous section, it is clear that one might improve the model by taking a displacement field closer to the hydrodynamic form.

We next need to determine the strength κ. A semi-empirical way to do this is to make use of the isospin-dependent optical potential. The single-particle potential depends on isospin approximately as

$$V_{n,p} = \tau_z V_\tau \frac{N - Z}{A} \quad .$$

with $V_\tau \approx 26$ MeV. Assuming the potential is caused by a short-range interaction, we can write for an arbitrary transition density

$$\delta V = V_\tau \rho_{sym} \tau_z / \rho_0 \quad .$$

Now let us choose κ in eq. (5.15) to be consistent with the above equation when applied to the collective transition density. We take $\rho_{sym} = cz$ and find

$$\delta V = V_\tau cz = \kappa z \int z' cz' \rho_0 d^3 r' \quad .$$

This may be solved for κ, to obtain

$$\kappa = \frac{3 V_\tau}{A < r^2 >} \quad .$$

We are now ready to calculate the dipole frequency. We obtain this from the dynamic polarizability in RPA, which diverges at frequencies ω that satisfy

$$\kappa \sum_{ph} \frac{2(e_h - e_h) <p|z\tau_z|h>^2}{(e_p - e_h)^2 - (\hbar\omega_D)^2} = -1 \ .$$

For a quick estimate, let us assume that all the relevant particle-hole energies are degenerate at the oscillator frequency ω_0, and evaluate the sum of the squared matrix elements using the TRK sum rule for noninteracting particles:

$$\sum_{ph} <p|z\tau_z|h>^2 \hbar\omega_0 = \frac{\hbar^2 A}{2M} \ .$$

The resulting equation for ω is

$$(\hbar\omega_D)^2 = (\hbar\omega_0)^2 + \frac{3\hbar^2 V_\tau}{M <r^2>} = \frac{1}{A^{2/3}} \left[(41)^2 + (61)^2\right] \ .$$

We see that the interaction contributes to the energy of the vibration more that the kinetic energy of single-particle motion. Evaluating the above relation gives

$$\hbar_D\omega \approx \frac{73}{A^{1/3}}.$$

This is close to the empirical dependence, but slightly lower. One effect that raises the frequency that has not been included is the velocity-dependence of the interaction. In effect, the single-particle energy estimated in the harmonic oscillator model, eq. (A.2), should be higher because the potential depends on the shell of the particle. It is beyond our scope to include such effects here[§]. We only mention that these interactions affect the TRK sum rule as well, increasing it by 15–20%.

[§] See Bohr and Mottelson (1975) for a discussion along lines similar to those we followed in the above derivation.

6

Surface modes

A liquid drop can sustain surface oscillations which do not compress the fluid. These are analogous to ripples on the surface of an infinite fluid; in both cases the restoring force is provided by the surface tension of the liquid. For a spherical drop of an ideal fluid, the displacement field for these oscillations is simply the gradient of a multipole field

$$\vec{u} = a(t)\vec{u}_0 = a(t)\nabla r^L Y_{LM}(\theta, \phi) \ , \tag{6.1}$$

to lowest order in the deformation parameter a (cf. eq. (2.7) and app. B). As mentioned in Chap. 2, this displacement field moves the fluid without compression, since its divergence vanishes, $\nabla \cdot \vec{u} = a\nabla^2 r^L Y_{LM} = 0$. Because the motion is described by a velocity potential, namely $r^l Y_{LM}$, the motion is also irrotational. Note that the lowest nontrivial multipole is $L = 2$. Eq. (6.1) vanishes for $L = 0$, since there is no way to make spherically symmetric irrotational motion without compression. We will defer until the next chapter an examination of compressional modes. The lowest multipole, $L = 1$, is the dipole mode. We saw in the last chapter that the dipole mode requires two kinds of particles moving opposite to each other; a uniform displacement of the whole system does not produce an excitation.

In quantum physics, the modes controlled by surface tension arise naturally in superfluids, such as nuclei and liquid ^{4}He at low temperature. We will consider in the next section the classical theory, and apply it to ^{4}He drops. However, the theory here is

99

somewhat of an academic exercise, because it is still experimentally quite difficult to study small helium drops[†].

The theory of surface modes has much greater application to nuclear excitations(cf. e.g. Bohr and Mottelson (1975)). The quadrupole mode, $L = 2$, appears in several different contexts. The equilibrium shape of some nuclei is deformed, mainly by quadrupole distortions. Then the excitations form a rotational band, with large quadrupole transition strengths between members of the band. Many nuclei, including deformed nuclei, have low-frequency quadrupole vibrations. These correspond most closely to the liquid drop vibrations, although there are important differences. We shall discuss these modes in Sec. 6.4 (cf. also Sec. 2.1). Finally, all but the lightest nuclei exhibit the giant quadrupole vibration. This mode has the same displacement field as the liquid drop oscillation, but the restoring force is better described in classical terms as arising from an elastic shear modulus (cf. Sec. 6.3 and 2.1).

Since the modes in Fermi systems such as nuclei are so much richer and therefore more complicated than in classical systems, we will also examine the idealized limit in which the system is infinitely large and the surface is flat in equilibrium. This is called the semi-infinite slab model; it is discussed in more detail in Sec. 6.2.

Before beginning the discussion of quantum systems, we briefly review the classical theory of ripples on a fluid with a flat surface. In Fig. 6.1 is sketched a standing wave on the surface of a fluid. We describe the motion with a coordinate system in which the equilibrium is surface in the xy plane at $z = 0$. The normal modes have a sinusoidal dependence on the xy coordinates, because of the translational invariance in the xy plane. Assuming the motion to be incompressible, as above, the divergence of the displacement field must vanish, and this restricts the form of the displacement field to be exponential in the z-direction. Let us assume the motion is along the x-direction with a reduced wavenumber q, writing the field as $\vec{u} = a(\vec{\nabla}\exp(qz + iqx))/q = a(\hat{z} + i\hat{x})\exp(qz + iqx)$. The change in energy of the liquid under this displacement is due to the increased surface area, which may be evaluated as (cf. Fig. 6.2)

[†] References to the experimental literature may be found in Scheidemann et al. (1990).

$$\mathscr{V}(A) = \int \gamma d\mathscr{A}$$

$$= \int \gamma \sqrt{1 + \left|\frac{du_z}{dx}\right|^2_{z=0}} \, dx \, dy$$

$$= \gamma\mathscr{A} + \frac{1}{2}\gamma\mathscr{A}q^2a^2 + \dots$$

where γ is the surface tension and $\mathscr{A}$ is the total area of the fluid. The associated restoring force constant for collective motion is (cf. eq. (2.11))

$$(C_{liq})_S = \frac{\partial^2 \mathscr{V}}{\partial a^2} = \gamma\mathscr{A}q^2,$$

where the labels *liq* and S stand for liquid and slab respectively. The inertia associated with the field is

$$(I_{inc})_S = \int_{-\infty}^{0} dz \int dx \, dy \, m\rho_0 |\nabla e^{(qz+iqx)}|^2 / q^2$$

$$= \frac{m\rho_0\mathscr{A}}{q} \, .$$

Then the Rayleigh principle gives the oscillation frequency

$$(\omega_{liq})_S^2 = \frac{\gamma q^3}{m\rho_0} \, . \tag{6.2}$$

As an example, we give the numbers for liquid ^{4}He, taking the reduced wave number to be $q = 0.1$ Å^{-1}. The physical quantities needed in the above formula have the values

$$\gamma = 2.4 \times 10^{-5} \text{ eV}, \tag{6.3}$$

$$\rho = 0.036 \text{ Å}^{-3} \, ,$$

$$\frac{\hbar^2}{m} = 1.04 \times 10^{-3} \text{ eV-Å}^2,$$

where m is the mass of a ^{4}He atom. Inserting this in eq. (6.2), we find an excitation energy for the mode of $\hbar\omega = 2.6 \times 10^{-5}$ eV, which corresponds to an ordinary frequency $f = 2\pi\omega = 2.5 \times 10^{11}$ Hz.

We have given a purely classical derivation, but a very similar formula may be derived for the quantum surface oscillations

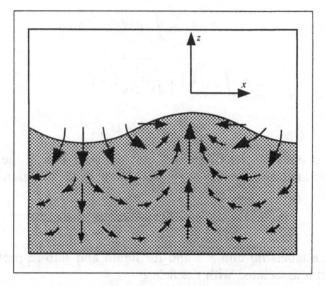

Fig. 6.1. Displacement field for a standing wave on the surface of a liquid.

of a Bose superfluid. The derivation uses the diabatic frequency formula, eq. (3.25), together with a mean field theory of the Hamiltonian (see Bertsch (1974)).

6.1   Liquid drop vibrations

We now go from semi-infinite to spherical geometry. As before, we assume that the flow is irrotational and incompressible, and therefore the displacement is given by eq. (6.1) to first order in the deformation parameter a. We again want to calculate the increase in surface area to find the surface energy associated with the distortion, but the calculation is more complicated in this case. The area of the drop is given by the formula,

$$\mathscr{A} = \int \frac{R^2(\theta, \phi)}{\cos \psi} d\Omega \; , \qquad (6.4)$$

where R is the distance to the surface along the radial direction (θ, ϕ), and ψ is the angle between the normal to the surface and the radial vector, as shown in Fig. 6.2. This expression must be evaluated to second order in a. Unlike other cases, it will be necessary to include a second-order term in the displacement field to preserve the initial volume. We expand R as

$$R(\theta, \phi) = R_0 + au_{0r}(R_0, \theta, \phi) + a^2c \ ,$$

where R_0 is the spherical radius and the parameter c controls the volume. We also adopt a notation denoting the radial component of a vector with the subscript r and the tangential component with t. The cosine factor in the surface area formula is given to second order by

$$\frac{1}{\cos \psi} = \sqrt{1 + (\nabla_t R)^2} \approx 1 + \frac{1}{2}|\nabla_t u_r(R_0, \theta, \phi)|^2 \ .$$

We can now evaluate eq. (6.4) to second order in a, obtaining

$$\mathscr{A} = 4\pi R_0^2 + 2\int d\Omega R_0 u_r + \frac{1}{2}R_0^2 \int |\nabla_t u_r|^2 d\Omega$$

$$+ 2R_0 c a^2 \int d\Omega + a^2 \int |u_{0r}|^2 d\Omega + \dots \ .$$

To evaluate these integrals, we first note that $u_{0r} = \nabla_r r^L Y_{LM} = Lr^{L-1}Y_{LM}(\theta, \phi)$. Then the first integral vanishes for $L \neq 0$, by the orthogonality of the spherical harmonics. The quantity $\nabla_t u_{0r} = Lr^{L-1}\nabla_t Y_{LM}$ enters in the argument of the third term. Consequently, this term is proportional to the angular integral $\int d\Omega |\nabla_t Y_{LM}|^2$. This is evaluated by integrating by parts, using the basic relation $\nabla_t^2 Y_{LM} = -L(L+1)Y_{LM}/r^2$ to obtain

$$\int d\Omega |\nabla_t Y_{LM}|^2 = L(L+1)/r^2 \ .$$

We next determine c by demanding a constant volume, $\int d\Omega R^3 = 4\pi R_0^3/3$. This yields $c = -(Lr^{L-1})^2/R_0$.

Before writing down the final formula for the distorted surface area, we make one change of notation. It is convenient to characterize the amplitude of the distortion by the root mean square distance the surface moves,

$$d_L = \left[\int |u_r|^2_{r=R_0} d\Omega \right]^{1/2} = aLR_0^{L-1} \int d\Omega |Y_{LM}|^2 = aLR_0^{L-1} \ .$$
$$(6.5)$$

The formula for the surface area is then

$$\mathscr{A} = 4\pi R_0^2 + \frac{1}{2}(L-1)(L+2)d_L^2 + \dots \ . \tag{6.6}$$

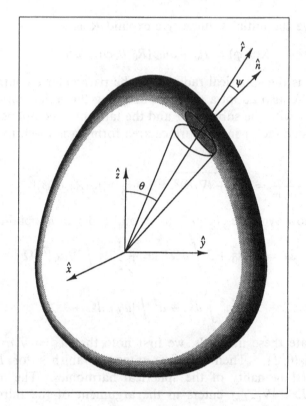

Fig. 6.2. Surface area element of liquid drop.

The numerator function that goes in the Rayleigh principle is given by

$$C_{liq,L}^d = \gamma \frac{d^2 \mathcal{A}}{d \, d_L^2} = \gamma(L-1)(L+2) \ . \qquad (6.7)$$

In terms of the quantity d_L, the displacement field in (6.1) becomes $\vec{u}_0 = \vec{\nabla} r^L Y_{LM} / (L R_0^{L-1})$. Making use of the relation (2.10) we can calculate the inertia associated with the motion. This is simply given by

$$I_{inc,L}^d = \frac{m\rho_0}{(L R_0^{L-1})^2} \int_0^{R_0} r^2 dr \int d\Omega (|u_{0t}|^2 + |u_{0r}|^2)$$

$$= m\rho_0 \int_0^{R_0} r^2 dr \frac{((L(L+1) + L^2) r^{2L-2}}{L^2 R_0^{2L-2}}$$

$$= \frac{m\rho_0 R_0^3}{L}.$$

The liquid drop vibrational frequency is just the ratio of the above expressions:

$$\omega_{liq,L}^2 = \frac{L(L-1)(L+2)\gamma}{m\rho_0 R_0^3}. \tag{6.8}$$

For quadrupole vibrations (L=2) this expression coincides with eq. (2.28). The difference between the inertia and restoring force derived above and those of eqs. (2.26) and (2.27) are simply due to the different parameterization of the deformation used[†].

Notice that the frequency vanishes for $L = 1$. Recall that $L = 1$ just represents a uniform displacement of the sphere, and the frequency vanishes here because there is no distortion of the liquid. In the limit of a very large drop, this equation must reduce to the dispersion formula for an infinite surface, eq. (6.2). The connection is easily made, replacing multipolarity L by the approximate wave number relation, $L = qR_0$.

As in the case of an infinite surface, a corresponding theory of quantum surface oscillations may be derived. One assumes that the ground state is a Bose superfluid and the interaction stabilizing the ground state is short-range. Droplets of liquid helium should therefore exhibit these surface waves below the superfluid transition temperature.

To get an idea of the excitation energies involved, we evaluate eq. (6.8) for the $L = 2$ mode in a ^{4}He cluster with 1000 atoms. Taking the other constants from eq. (6.3), we find the excitation energy to be $\hbar\omega = 2.8 \times 10^{-5}$ eV $= 0.33$ K.[‡] This is well below the excitation energy of bulk modes such as phonons or the roton. Of course, it would be extremely difficult to detect this oscillation experimentally. It would not couple directly to dipole photons, but could be created by the inelastic scattering of photons. This is still possible in principle; such modes are called Raman-active.

[†] To transform from one coordinate to the other, use $d^2/d(d_2)^2 = \left(d\epsilon/d(d_2)\right)^2 d^2/d\epsilon^2 = (5/16\pi R_0^2)d^2/d\epsilon^2$. This gives $I^\epsilon = R_0^2(16\pi/5)I^d$.

[‡] The kelvin unit of energy, 1 K $= 0.864 \times 10^{-4}$ eV, is commonly used in the physics of liquid helium.

6.2   Surface oscillations of Fermi liquids

The behavior of Fermi systems is quite different from Bose su-perfluids, in fact enough so that the usual designation as a Fermi "liquid" is somewhat misleading. Fermi systems have a huge density of states at low excitation energy compared to Bose systems, and the motion tends to be quite dissipative. In cold Bose systems the single-particle motion is governed by a unique wave function, that of the condensate, but in the mean field theory of a fermion system all particles have different wave functions, as required by the Pauli principle, and coherence between them tends to quickly die out. This is the reason why classical mechanics applies well to Bose systems and less to Fermi systems.

Before studying the surface oscillations in finite systems, it is useful to see the general properties of the semi-infinite Fermi system having a free surface. This can be realized in nature in liquid ^{3}He, or as a limiting behavior of very large nuclei. We consider a Fermi system having the same geometry as in Fig. 6.1. The potential well in a mean-field model will be attractive, going to a constant value for large negative z, which will bind the particles. For large positive z, the potential goes to zero. The single-particle wave functions are plane waves in the x- and y-directions, but their z-dependence can only be found by solving the one-dimensional Schrödinger equation

$$-\frac{\hbar^2}{2m}\frac{d^2}{dz^2}\phi_k(z) + V(z)\phi_k(z) = \epsilon_z\phi_k(z) \ .$$

Of course in mean field theory the potential $V(z)$ must be self-consistent in that the particle interaction with the many-particle wave function generates the same potential as is used to determine the single-particle wave functions.

We now describe a surface oscillation of the system. Since the interior density does not change, the potential will be the same for large negative z. But if the surface moves a distance δz, the self-consistent potential must move the same amount. Thus, there will be an oscillating potential that depends on z as $V(z-\delta z)$. The dependence in the x and y directions is sinusoidal, so we can write the spatial dependence of the oscillating part of the potential as

$$\delta V(r) = \delta z \frac{dV}{dz}\sin k_t \cdot r_t \ . \tag{6.9}$$

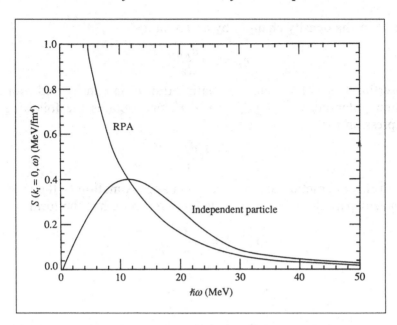

Fig. 6.3. Surface response function for a semi-infinite Fermi system, from Esbensen and Bertsch (1984).

The first step is to calculate the independent particle response to this field. This was done for semi-infinite nuclear matter by Esbensen and Bertsch (1984). The resulting imaginary part of the polarization is shown in Fig. 6.3, for $k_t \ll k_F$. The function is rather smooth and broad, because there are no shells in a semi-infinite system to produce structure. It is quite well described by a simple Lorentzian function,

$$\Pi^0 \approx A(k_t)\left(\frac{1}{\omega_0(k_t) - \omega - i\gamma(k_t)} + \frac{1}{\omega_0(k_t) + \omega + i\gamma(k_t)}\right) \,,$$

provided the surface wavenumber k_t is not too high.

Next we calculate the RPA response, making use of a separable interaction, i.e. one of the type eq. (4.13),

$$v(r, r') = \kappa f(r)f(r') \quad .$$

We take f of the form eq. (6.9), so the transition potential will automatically have that form. The coefficient κ is determined by a self-consistency argument. Namely, if the surface moves a distance

δz, then the density changes by an amount

$$\delta\rho = \delta z \frac{d\rho_0}{dz} \ .$$

Inserting this in the equation relating transition density and transition potential, $\delta V = \int \delta\rho v(r,r')d^3r'$, one obtains the following expression for κ:

$$\frac{1}{\kappa} = \int d^3r \frac{d\rho_0}{dz}\frac{dV}{dz}.$$

With this interaction and a Lorentzian approximation to the independent particle polarization, the RPA response has the form

$$\Pi^{RPA} = \frac{\Pi^0}{1 - \kappa\Pi^0}$$

$$\sim \frac{1}{\omega^2 + ib\omega + \omega_0^2(k_t)}.$$

Here $\omega_0^2(k_t)$ is a function of the surface wave number k_t, depending on the interaction strength κ as well. The RPA response for vanishing k_t is shown in Fig. 6.3. We see that it is enhanced over the free response at low frequencies, as is to be expected for an attractive interaction. The response actually diverges at $k_t = 0$, due to the vanishing of ω_0^2. With self-consistency imposed on the interaction strength κ, ω_0^2 is proportional to k_t^2 to lowest order, behaving as a surface energy. The divergence of the response at $k_t = 0$ has a simple physical explanation. The $k_t = 0$ response represents nothing more than a shift of the surface position, which costs no energy at all. Therefore, any perturbation at $k_t = 0$ will produce an infinite response[†].

The small k_t and ω behavior of the polarization propagator thus has the form

$$\Pi^{RPA} \sim \frac{1}{ib\omega + \gamma k_t^2}. \tag{6.10}$$

This is exactly the response associated with the diffusion equation, in which the displacement of the surface position u_z satisfies

$$b\frac{\partial u_z}{\partial t} = \gamma\nabla_t^2 u_z \ .$$

---

† In quantum field theory, such zero-energy modes are called Goldstone bosons.

The coefficient b can furthermore be calculated by a very simple model, called the piston model or the **wall formula** (Blocki et al (1978)). In this model, the free surface is replaced by an external wall, and the energy dissipation associated with uniform motion of the wall is calculated classically. The energy dissipation rate is linearly proportional to the velocity of the wall, implying a linear friction, as described in Chap. 2. The formula for the frictional force per unit area, $F/\mathscr{A}$, is (Bertsch and Esbensen (1985))

$$F/\mathscr{A} = bv = \frac{3}{4}\rho_0 p_f v \ ,$$

where $p_f = \hbar k_f$ is the Fermi momentum and v is the velocity of the wall. It is remarkable that the self-consistent RPA agrees with a manifestly nonconsistent classical calculation. However, the concept of friction has limited validity for small systems; this will be discussed at greater length in Chap. 9.

Of course, the full RPA response is much more subtle. We will see that the density of particle-hole states plays a decisive role in determining the collectivity of the response and such questions as whether or not a low-frequency mode exists. Eq. (6.10) should be viewed as a qualitative formula to give the average response at low frequency and long wavelength.

6.3 Nuclear quadrupole modes

As mentioned in the introduction, there are several distinct modes of quadrupolar motion in nuclei. A few definitions are needed for a quantitative discussion. The quadrupole moment of a system is conventionally defined as the integral of the operator $Q = 2z^2 - x^2 - y^2$ over the charge density of the system. However, for discussing electromagnetic transitions, a more convenient quantity is the $B(E2)$, defined by

$$B(E2; i \to f) = \sum_{fM} | <i|er^2 Y_{2M}|f> |^2 \ . \tag{6.11}$$

where the sum is over magnetic substates of the final state f, and the multipole operator acts only on the protons. Note that in this definition it is necessary to specify which is the initial and final state. However, the definition has the advantage that the electromagnetic decay rate can be expressed directly in terms of the $B(E2)$:

$$T(E2; I_i \to I_f) = W = \Gamma/\hbar = 1.22 \times 10^9 E^5 B(E2; I_i \to I_f) \ \text{s}^{-1}$$

where E is in MeV and the $B(E2)$ is in units of e^2 fm^4. A typical example of a quadrupole transition is the decay of the lowest 2^+ excited state of ^{208}Pb. Here the various quantities have the following values[†], the last one of which can be verified from the above formula,

$$B(E2; 2 \to 0) = 0.6 \times 10^3 \ e^2 \ \text{fm}^4; \quad E = 4.07 \ \text{MeV};$$

$$T(E2; 2 \to 0) = 8 \times 10^{14} \ \text{s}^{-1}; \quad \Gamma = 0.54 \ \text{eV} \ .$$

The energy-weighted sum rule for the $B(E2)$ may be evaluated from eq. (3.11). The result for an angular momentum zero ground state is

$$\sum_f B(E2; f \to 0)(E_f - E_0) = \frac{5}{4\pi} \frac{\hbar^2}{m} Z e^2 <r^2> \ . \tag{6.12}$$

For example, the sum rule value for the nucleus ^{208}Pb has the approximate value 0.6×10^5 MeV^2 fm^5. We obtained this using for the expectation value of r^2 the estimate $<r^2> \approx 3(1.2A^{1/3})^2/5$ fm^2. Thus the lowest 2+ state of ^{208}Pb carries approximately 6 % of the energy-weighted sum rule.

We can also define a sum rule for the mass quadrupole operator instead of the charge; the only effect is to replace the $e^2 Z$ by A. Another variant of the sum rule is the sum for the charge operator, but only including transitions in which the neutrons and protons move together. This is the isoscalar sum rule; in nuclei with equal numbers of neutrons and protons, this will be half the full charge sum rule. The state at 4.07 MeV in ^{208}Pb is of this type and consequently it will exhaust about 2×6 % of the isoscalar sum rule (cf. e.g. Table 6.8 of Bohr and Mottelson (1975)).

When the vibration is excited by strong interactions, one is mainly sensitive to the surface motion, and it is convenient to describe the vibration within the framework of a collective model. For example, in the Tassie model, eq. (3.10), the shape of the transition density is completely specified, and a transition can be characterized by a single parameter, its amplitude. We write this

---

[†] The $B(E2)$ values of the lowest transitions are compiled by Raman et al. (1987).

in the form

$$<LM|\hat{\rho}|0> = \frac{d_L}{\sqrt{2L+1}}(r/R_0)^{L-1}\frac{d\rho_0}{dr}Y_{LM}^*(\hat{r}) \ ,$$

where R_0 is the nuclear radius and the amplitude d_L is called the **deformation length**[†]. If we drop the factor $(r/R_0)^{L-1}$ in the above formula, we obtain the collective model of Bohr and Mottelson. They define a dimensionless deformation parameter β_L related to d_L by (cf. App. B)

$$\beta_L R_0 = d_L,$$

giving a parameterization

$$<LM|\hat{\rho}|0> = \frac{\beta_L R_0}{\sqrt{2L+1}}\frac{d\rho_0}{dr}Y_{LM}^*(\hat{r}) \ . \tag{6.13}$$

The $B(EL)$ associated with this transition density may be obtained from the definition, and is given by

$$B(EL; L \to 0) = \left(\frac{3}{4\pi}ZeR^L\right)^2\beta_L^2 \ .$$

One can also derive a sum rule for the deformation lengths, assuming that all transitions can be described by eq. (6.13) (Satchler (1971)). The sum rule reads

$$\sum_n (\beta_L R_0)_n^2(E_n - E_0) = 2\pi\frac{L(2L+1)}{3Am} \ .$$

With these preliminaries we discuss the two prominent quadrupole oscillations, namely the giant quadrupole vibration and the low-lying vibration.

Giant quadrupole

Nuclei heavier than about $A = 20$ systematically show a quadrupole resonance that has about 75% of the isoscalar sum rule. Thus, we may consider this a collective oscillation of most of the nucleons in the nucleus. The systematics of measurements by inelastic scattering (Bertrand (1976)) are shown in Fig. (6.4). The frequency of the oscillation is approximately given by

$$\hbar\omega = 63/A^{1/3} \text{ MeV} \ .$$

[†] Note the similarity, except for the factor $\sqrt{2L+1}$, to the classical d_L defined above eq. (6.5).

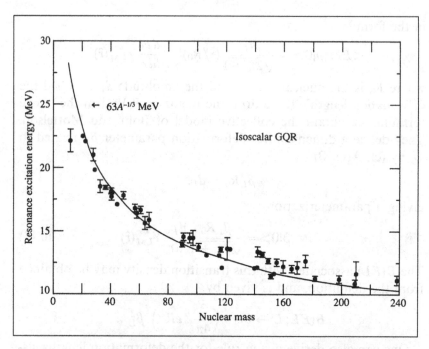

Fig. 6.4. Systematics of the nuclear giant quadrupole vibration.

A typical frequency of the oscillation is 10.5 MeV in ^{208}Pb. The resonance has a width of a few MeV, so in classical terms the vibration oscillates a few cycles before decaying. Obviously, this behavior is quite different from the strongly dissipative motion in the semi-infinite system.

From the experimental results above, it is clear that the dynamics associated with the giant quadrupole resonance is very different from that described by the liquid drop model. In fact, the liquid drop formula for the frequency, eq. (6.8) or equivalently (2.28), gives a different A-dependence. This is estimated as

$$\hbar\omega_{liq,Q} = \sqrt{32\pi\gamma\hbar^2/3mA}$$

$$\approx \sqrt{\frac{32\pi(1\ \mathrm{MeV/fm^2})(197.3\ \mathrm{MeV\ fm/c})^2}{(931\ \mathrm{MeV/c^2})(208)}}$$

$$\approx \frac{37\mathrm{MeV}}{\sqrt{A}}.$$

Furthermore, it gives numerical values which are nearly an order of magnitude smaller than those experimentally observed. For example, for ^{208}Pb the formula gives $\hbar\omega \approx 2.6$ MeV.

The reason the liquid drop model fails in the present case is that the restoring force eq. (6.7) only applies to slow shape changes in a fermion system. This condition is fulfilled more or less for the low-lying nuclear vibrations, but not for the giant quadrupole. On the other hand, RPA calculations give a rather good account of this vibration, as will be described in more detail later. Since we are interested in the properties of a collective oscillation at high frequency, the diabatic formula, eq. (3.25), is more applicable. Recall that the denominator in that formula was just the classical inertia for the multipole field, and the numerator plays the role of the potential energy function in classical physics. When one studies the expectation value in the numerator, one finds that the motion essentially commutes with the potential field, because the potential field moves with the particles. Of course, this is not exactly true for finite range interactions, which give an explicit contribution to the surface energy. But as we have seen, one does not get a large enough restoring force from the liquid drop component. The other commutator in the numerator is with the single-particle kinetic energy operator. Here there is a rather subtle effect, with the quadrupolar motion increasing the single-particle kinetic energy.

If the collective motion has a shear component, that is, if the medium is expanded in one direction while shrunk in another direction, the wavelengths of the single-particle wave functions will change in the same way, and the kinetic energy of the particles will be altered. To first order there is no change in the total kinetic energy for a shear without compression. But in second order the kinetic energy increases. This may be seen easily in a quantitative way for the quadrupolar distortion, treated with a scaling field. To preserve the density to all orders, we consider a transformation of the x, y, z coordinates as in eq. (2.25),

$$x' = xe^{-\epsilon}, \quad y' = ye^{-\epsilon}, \quad z' = ze^{2\epsilon} . \tag{6.14}$$

We treat the single-particle kinetic energy in a Fermi gas model. The energy per particle is

$$K_0 = \frac{1}{2m}\left(<p_x^2> + <p_y^2> + <p_z^2> \right) = \frac{3p_f^2}{10m}$$

in the undeformed system, where the relation $<p^2> = \frac{3}{5}p_f^2$ has been used. Under the deformation, the expectation values of the momenta change inversely as the coordinates:

$$<p_{x'}^2>=<p_x^2> e^{2\epsilon}, \quad <p_{y'}^2>=<p_y^2> e^{2\epsilon}, \quad <p_{z'}^2>=<p_z^2> e^{-4\epsilon} .$$

Inserting this into the kinetic energy formula, and expanding to second order in ϵ, we find

$$K(\epsilon) = K_0 + \Delta K$$

where

$$\Delta K = \frac{6\hbar^2 k_f^2}{5m}\epsilon^2 .$$

With this mode of distorting the system, the change in energy of the system is $N\Delta K$ which can be a very large change in energy for a small distortion, because the effect is proportional to the number of particles in the system. To produce a frequency formula for this, we need also the inertia which is given by eq. (2.26). The effective restoring force[†] for this collective energy is $d^2\Delta K/d\epsilon^2 = 12\hbar^2 k_f^2/5m$. The ratio of this force constant to the inertia gives the frequency formula

$$\omega_{el}^2 = \frac{6\hbar^2 k_f^2}{5m^2 <r^2>}. \tag{6.15}$$

With the usual nuclear parameters, $k_f = 1.34$ fm^{-1} and $<r^2> = 0.6(1.2A^{1/3})^2$ fm^2, the formula predicts $\hbar\omega_{el} \approx 64/A^{1/3}$ MeV. This has the correct functional form and also is numerically very close to the empirical value.

The above result can be derived from the diabatic frequency formula, but the algebra is too complicated to go through here. When the commutator of the kinetic energy with the displacement field is evaluated in eq. (3.25), one of the terms is

$$\int d^3r \sum_{i,j}(\nabla_i u_j)^2 \sum_l |\nabla_i \phi_l|^2 ,$$

where ϕ_l is a single-particle wave function. This function has just the same dependence on the displacement field as the shear energy in the theory of elasticity, given by the second term in eq. (2.29). Thus, for diabatic motion the Fermi liquid has a shear modulus.

[†] In terms of the more usual coordinate β_2 instead of ϵ, the restoring force parameter is $0.5N\epsilon_F$ (cf. App. B).

The sum over single-particle wave functions in the above equation may be evaluated in the Fermi gas model, to give the following formula for the shear modulus μ:

$$\mu = \frac{\hbar^2 k_f^2 \rho_0}{5m} .$$

Inserting this shear modulus into the variational estimate of the classical quadrupolar vibration, eq. (2.31), gives the same result as eq. (6.15).

We have not yet answered the question of why the giant quadrupole mode is a well-defined resonance and not just a highly damped broad continuum. The quadrupole mode is unique in that its shear field is constant over the entire system. The scaling transformation, eq. (6.14), distorts the Fermi surface the same amount everywhere. The particles in a Fermi system lose coherence by moving from one point to another where the collective field is different. If the Fermi surface were distorted in different ways at different points, the motion of the particles would eventually blur its sharpness.

In RPA, collectivity occurs when there is a strong residual interaction, and the collective strength is pushed into a frequency domain lacking single-particle excitations. We can see how this works in the RPA calculation of the quadrupole response function of ^{40}Ca. This is done in a discrete basis of particle states, to make the contributions of various particle-hole states distinct. In the top part of Fig. 6.5 we see the independent particle response. Most of the strength in Fig. 6.5 is at ~ 24 MeV excitation, corresponding to transitions to higher shells. The most important are the high-L orbitals, making a transition to a state with $L' = L + 2$. The corresponding RPA response function is shown in the lower part of the figure. The calculations were carried out within the framework of the Hartree–Fock approximation, making use of a density-dependent Skyrme interaction (cf. Sect. 4.1). This attractive interaction causes a collective state to appear at lower energy, ~ 18MeV. This follows the qualitative behavior of Fig. 4.1. Since the level density is low in this part of the spectrum, the collective strength is concentrated in a few nearby states.

We now present a rough analytic calculation of the giant quadrupole mode using a harmonic oscillator single-particle Hamiltonian and a separable interaction. The separable interaction is taken to

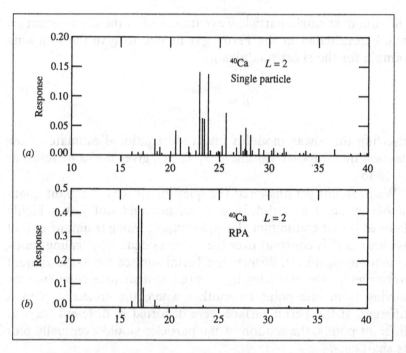

Fig. 6.5. Quadrupole response in ^{40}Ca: (a) independent particle; (b) RPA.

be of the form

$$v(r,r') = \sum \kappa_L F_{LM}(\hat{r}) F_{LM}^*(\hat{r}')$$

with

$$F_{LM}(r) = r^L Y_{LM}(\hat{r}) \ .$$

We first derive the self-consistent coupling constant κ of Bohr and Mottelson (1975). In an oscillation governed by this interaction, the transition potential will be proportional to F. We next assume that the oscillation is completely collective and use the equation of continuity to determine the form of the transition density. This yields the Tassie transition density,

$$\delta\rho = \vec{u} \cdot \nabla\rho_0(r)$$

where the displacement field is proportional to ∇F,

$$\vec{u}(r) = a(t)\nabla F_{LM} \ .$$

The transition potential must be consistent with this displacement field, which means that it must be derivable from the static potential by displacement. In the present model the static potential is the harmonic oscillator potential, $V(r) = \frac{1}{2}m\omega_0^2 r^2$, and so the self-consistent transition potential is

$$\delta V = \vec{u} \cdot \nabla V = am\omega_0^2 \vec{r} \cdot \nabla F_{LM}$$

$$= am\omega^2 L F_{LM} \ .$$

We now demand that the separable interaction gives the same δV, when convoluted with the transition density. This requires that

$$\kappa_L F_{LM}(r) \int d^3r' F_{LM}^*(r')\delta\rho = La(t)m\omega_0^2 F_{LM}(r) \ .$$

Note $F(r)$ cancels on both sides, showing the consistency in the choice of the shape of the field. The integral on the left hand side can be simplified using the identities on p. 102 as follows,

$$\int d^3r' F^* A\nabla F \cdot \nabla\rho_0 = -\int d^3r |\nabla F|^2 \rho_0$$

$$= -L(2L+1) \int r^2 dr r^{2L-2} \rho_0 \ .$$

This can be simplified and solved for κ to obtain

$$\kappa_L = -\frac{4\pi m\omega_0^2}{(2L+1) <r^{2L-2}>} \ ,$$

which for $L = 2$ is

$$\kappa_2 = -\frac{4\pi}{5} \frac{m\omega_0^2}{A <r^2>} \ .$$

Having found κ, we are now ready to solve the RPA dispersion relation eq. (4.15), which reads

$$1 = \kappa \sum_{ph} \frac{2(e_p - e_h) <0|F|ph>^2}{(e_p - e_h)^2 - (\hbar\omega^2)} \ .$$

The giant quadrupole excitations are produced by promoting a particle from one shell to a higher shell with principle quantum number two units higher. We consider here only nuclei with closed major shells, so there are no transitions within the same major shell. Transitions changing the principle quantum number by one unit are forbidden by parity. Thus the particle-hole energy in the above dispersion relation will be given by $e_p - e_h = 2\hbar\omega_0$. Assuming these energies all equal, the sum over particles and holes can be done in the numerator. This can be evaluated exactly from the mass sum rule, eq. (6.12) with the factor e^2Z replaced by A. Many factors cancel and one obtains a marvelously simple formula for the giant quadrupole frequency (Suzuki (1973),

$$\hbar\omega_Q = \sqrt{(2\hbar\omega_0)^2 - 2(\hbar\omega)^2} = \sqrt{2}\hbar\omega_0 \ . \qquad (6.16)$$

With the usual harmonic oscillator constant from eq. (A.2), the result is $\hbar\omega_Q = 58/A^{1/3}$, close to the empirical formula.

Eq. (6.16) has a simple interpretation. In the absence of the residual interaction, the restoring force would just be the extra energy of a particle in the oscillator potential, which is $2\hbar\omega_0$ for a quadrupolar excitation. To see the effect of the self-consistent potential, recall that the harmonic oscillator energy is half kinetic and half potential. By moving the potential with the particle, we keep the same potential energy. Thus the restoring force comes only from the single-particle kinetic energy, and is reduced by a factor of two from the noninteracting case. This gives a reduction in the frequency by the square root of two.

The above schematic calculation and the example with ^{40}Ca were particularly simple because the shell filling does not allow any quadrupole transitions at low excitation. The shell model for heavier nuclei gives some strength at low frequency, either because the shells are open, or because of the spin-orbit shell splitting. An example is the nucleus ^{208}Pb. Here there is a small amount of strength at about 5 MeV excitation, due to spin-orbit partners in certain shells, namely $l = 5$ and 6. For these shells, the higher j value is occupied while the lower j is empty. The excitation energy is just the spin-orbit energy splitting of the shell. The strength may be seen in Fig. 6.6. Including the residual interaction in RPA, we see that in addition to the giant resonance at 11 MeV, there is a large concentration of $B(E2)$ strength at 5.5 MeV. This excitation has mainly the structure of a particle with high orbital angular momentum move from one j-state to its spin-orbit coupled partner, but the $B(E2)$ is much larger than a pure single-particle excitation of this type would have. The excitation has in effect extracted some of the strength of the giant quadrupole, and carries 12 % of the energy-weighted sum rule, eq. (6.12). The RPA transition density for the lower quadrupole mode is shown in Fig. 6.7 together with the collective transition density, eq. (3.10) with

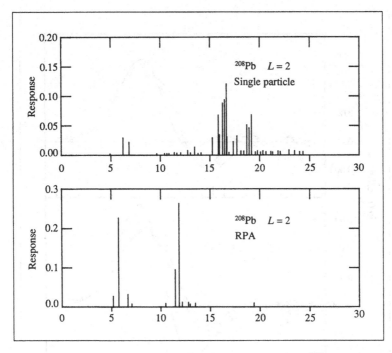

Fig. 6.6.   Quadrupole transition strength in ^{208}Pb

radial dependence

$$\delta\rho(r) \sim r\frac{d\rho_0}{dr} \ .$$

We see that the surface peaking dominates the transition density, but that the RPA density is more surface peaked than the collective model predicts.

6.4   Low frequency vibrations

In open shell nuclei, there is usually a quadrupole excitation at 1 or 2 MeV excitation energy. In terms of the sum rule, it does not have a large strength—typically 5-10% of the isoscalar sum rule. On the other hand, the $B(E2)$ is much larger than a single-particle transition in the independent particle shell model. Since this excitation is a common feature of the spectrum, we would like to understand it in collective terms.

For a description of the low part of the spectrum, the independent particle shell model completely breaks down. One must at

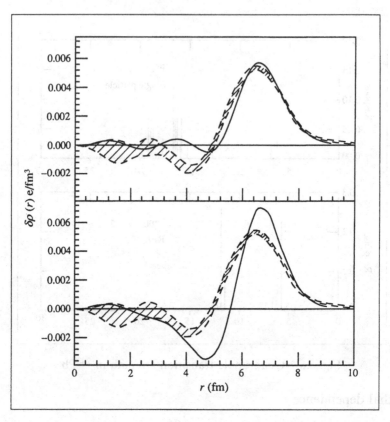

Fig. 6.7. Transition density of the low 2^+ excitation in ^{208}Pb, comparing theory and experiment. The collective model transition density is shown as the solid line in the upper figure, normalized to the experimental $B(E2)$. The measured transition density from electron shadowing is shown together with the experimental uncertainly as the hatched area. The lower figure shows the RPA transition density as the solid line.

least include the pairing interaction in the dynamics of the system to be at all useful. In fact one of the original applications of RPA to nuclear physics was the treatment of the quadrupole interaction in open shell nuclei described by pairing. The corresponding theory is called quasi-particle RPA; used with a separable interaction it was one of the early successes of RPA in nuclear physics (Bes and Sorensen (1969)). If the pairing is strong enough, the Fermi system will behave like a superfluid, and the liquid drop formula for the quadrupole frequency would apply. This would happen if the pairing gap were larger than the giant quadrupole frequency.

Since the pairing gap is only 1 MeV, we are far from this regime in ordinary nuclei. One would have to make nuclei with masses $A > 200,000$ to bring the giant quadrupole that low in frequency, and assume that the pairing gap remains constant.

Thus nuclei are in an intermediate regime where the pairing produces effects too large to ignore but too small to make the system completely superfluid. Fortunately, techniques exist to deal with motion in this regime, see e.g. Ring and Schuck (1980). The essential physics can be seen in the RPA, in which the particles are replaced by quasiparticles of pairing theory. To see qualitatively how this behaves, let us write once more the RPA dispersion relation with a separable interaction, but this time starting from a paired ground state. The dispersion relation reads

$$1 = \kappa \sum_{i,j} \frac{2(E_i + E_j) <0|f|\tilde{i}\tilde{j}>^2}{(E_i + E_j)^2 - \omega^2}$$

where $\tilde{i}, \tilde{j}$ label quasiparticles and $E_{i,j}$ are their energies[†] For our purposes here we only need know that the energies are all positive and have a minimum value of the gap constant Δ. We are now looking for a collective excitation of low frequency, so we will expand this equation in powers of ω^2, which requires that $\omega^2 << 4\Delta^2$. Truncating at the lowest nontrivial order, we obtain the following formula for the frequency:

$$\omega_{pl}^2 = \frac{1 - 2\kappa \sum_{ij} <0|f|ij>^2 / (E_i + E_j)}{2\kappa \sum_{ij} <0|f|ij>^2 / (E_i + E_j)^3} .$$

We wish to express this in the classical form, $\omega^2 = C/I$. Comparing with the stiffness coefficient (4.21) associated with the RPA polarizability calculated in Chap. 4,

$$C = \frac{1 - 2\kappa \sum_{ij} <0|f|ij>^2 / (E_i + E_j)}{\sum_{ij} <0|f|ij>^2 / (E_i + E_j)} ,$$

we obtain

$$\frac{1}{I} = \frac{\sum_{ij} <0|f|ij>^2 / (E_i + E_j)}{2\kappa \sum_{ij} <0|f|ij>^2 / (E_i + E_j)^3}.$$

[†] For completeness, we mention that the quasiparticle matrix elements are related to particle matrix elements by the Bardeen–Cooper–Schrieffer formula, $<0|f|\tilde{i}\tilde{j}> = <i|f|j> (U_i V_j + V_i U_j)$.

In the formula for the inertia, the numerator and denominator have averages of $(E_i + E_j)^{-1}$ and $(E_i + E_j)^{-3}$ respectively. For a crude estimate of the inertia we simply replace the $E_{i,j}$ by the pairing gap Δ and find

$$I \sim \frac{1}{\Delta^2} .$$

This quadratic dependence of the collective inertia on the pairing gap is a common feature of many models of low-frequency collective motion.

It is possible to extract this dependence on Δ directly from the pair-correlated wave functions, without making the quasiparticle transformation. The inertia relates the collective kinetic energy to the collective momentum of the wavefunction, which in turn depends on the change in phase of the wavefunction as the collective coordinate is varied. Representing the full wavefunction by a linear combination of Hartree–Fock configurations, we ask how the energy changes as the relative phase of nearby configurations is varied. In the pairing theory, there is some interaction matrix element connecting states in which a pair of particles has jumped from one orbital to another. Let us call this strength g. With the pairing coherence, the many particle-state has amplitudes in many orbitals for the pair to jump to, and the total matrix element is increased from g to Δ. The pairing coherence also means that many pairs could jump, and produce the same final state. The number of pairs that can participate is proportional to Δ/g. The product of these two factors is the Hamiltonian matrix element connecting two nearby states. Thus the inertia scales as

$$I \sim \frac{g}{\Delta^2} .$$

When this pair-hopping model is calculated in detail (Bertsch et al (1986)), one finds that the inertia for quadrupolar motion in heavy nuclei is about ten times the classical inertia, e.g. eq. (2.26). This means that the motion described will only have about a tenth of the energy-weighted sum rule, which is the order of magnitude of what is found at low excitation energy.

The empirical energy of the lowest quadrupole excitation is shown in Fig. 6.8, for a variety of spherical nuclei. We see a tremendous fluctuation, which is due essentially to shell effects. At shell closures the particle-hole energy differences are larger and the quadrupole excitation is correspondingly high. Between

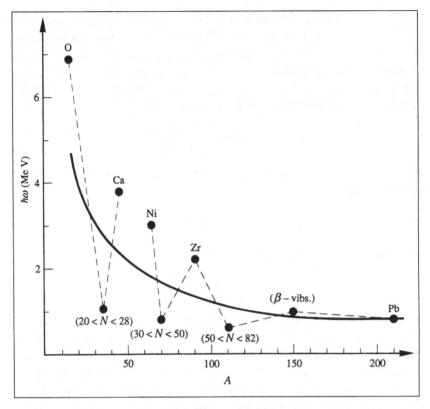

Fig. 6.8. Systematics of low quadrupole excitations.

closed shell, the particle-hole gap is small or may vanish and the quadrupole excitation comes down to the 1–2 MeV region. It is obviously not possible to describe this complex behavior in terms of global vibrational formulas with macroscopic quantities such as surface tension; etc. Nevertheless it is interesting to apply such models to see whether the average behavior can be described, with fluctuation effects removed.

An expression for the low-frequency inertia can be derived in the pair-hopping model[†] which has a simple dependence on the mass A (Bertsch (1988)). Putting in empirical values for the pairing strength, one finds

$$I^{\beta}_{pair} = \frac{\hbar^2}{175} A^2 \quad \text{MeV} \quad .$$

[†] The present application may be found in Broglia et al. (1994).

This inertia is with respect to the coordinate β_2 rather than ϵ, as we had in eq. (2.26), or d_L, as we have in eq. (6.7). We can examine in a very rough way the systematics of the low quadrupole state using this inertia together with the liquid drop restoring force. For heavy nuclei, the Coulomb energy modifies the force constant significantly, to give(Bohr and Mottelson (1975))

$$C_{liq,Q}^{\beta} = 4\gamma R_0^2 - \frac{3}{10\pi}\frac{Z^2 e^2}{R_c} \ .$$

where $R_c = 1.25 A^{1/3}$ fm is the Coulomb radius. Numerically, this is approximately

$$C_{liq,Q}^{\beta} = 6\left(1 - \frac{Z^2}{50A}\right)A^{2/3} \ .$$

The frequency is then given by $\sqrt{C/I}$, i.e.

$$\hbar\omega_{liq,Q}^{\beta} \approx \frac{30}{A^{2/3}}\left(1 - \frac{Z^2}{100A}\right) \ .$$

This is shown in Fig. 6.8 as the heavy line. We see that in some sense it does reproduce the average trend of the data. While there are no liquid drop modes for individual nuclei, the low quadrupole frequency of an "averaged" nucleus can be regarded in these terms. We see that there are very large fluctuations about this curve. These are associated with shell effects, which greatly affect the amount of pairing. However, on average the curve reproduces the trend of the data.

6.5   Higher multipolarities

In nuclear physics, the giant modes higher than the quadrupole do not have enough collectivity to be visible as a clear peak in a spectrum. The high energy modes are very much subject to Landau damping their presence can only be inferred by a careful analysis of inelastic scattering angular distributions (Speth and van der Woude (1981)).

However, in the low-energy part of the spectrum, there is often n very collective octupole state similar to the collective quadrupole state. It has only a small fraction of the energy-weighted sum rule, but its $B(E3)$ strength is enhanced over the single-particle values by an order of magnitude. Again, the RPA works well in closed shell nuclei to describe this mode. A well-studied example is in the nucleus ^{208}Pb, which has as its first excited state at 2.6 MeV a 3^- excitation. The octupole transition strength is 7% of the energy-weighted sum rule, eq. (3.12). Fig. 6.9 shows the transition density for this mode measured by electron scattering. The surface peaking of the transition density is very prominent, showing the strong collectivity. The peak in the transition density is at about 6.7 fm, which is close to the point where the ground state charge density varies most rapidly (see Fig. A.1). This is just what one would expect for the collective model transition density, eq. (6.13). In fact we can extract a value for the deformation length by comparing the peak transition density with the slope of the ground state density. One obtains

$$d_3 = \frac{\sqrt{7}\delta\rho_{max}}{\rho_0/4a_0} \approx \frac{\sqrt{7}(0.0075 \text{ e fm}^{-3})}{(0.025 \text{ e fm}^{-4})} = 0.8 \text{ fm} \quad,$$

where we have used a Fermi function parameterization of the ground state charge density distribution,

$$\rho(r) = \rho_0/(1 + \exp((r - R)/a_0)),$$

with $\rho_0 \approx 0.065$ e fm^{-3} (cf. Fig. A1). A nonvanishing transition density is also apparent in the interior. This is very much smaller and shows some oscillation. It is due to specific shell effects associated with the lowest particle-hole transitions. The figure also shows the predictions of a number of RPA calculations, differing on details of the single-particle Hamiltonian and the residual interaction. As long as these are chosen in some reasonable way, the results are similar. Not only the magnitude of the collectivity but the trend of the interior oscillation is reproduced.

This mode is very easily excited in the scattering of strongly interacting particles as well as electrons. In Fig. 1.12 we showed some angular distributions for inelastic proton scattering, which showed the characteristic diffraction pattern of a surface reaction. The figure also shows theoretical fits to the data, using the distorted-wave Born approximation to model the scattering wave function, and using the collective model for the transition potential. The deformation length extracted from this was $d_3 = 0.75$ fm, which is quite consistent with the electromagnetic transition density. An approximate formula for the cross section is derived in App. D, leading to

$$\sigma = \frac{d_L^2 R_S}{8a_0},$$

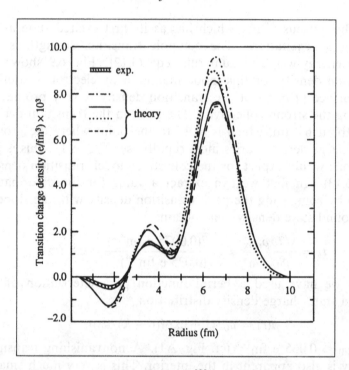

Fig. 6.9.   Octupole transition density in ^{208}Pb. Shaded curve: experiment; other curves: various RPA models (Heisenberg (1981)).

where $R_S = R + a_0 \approx 1.2A^{1/3} + 0.65$ fm is the strong absorption radius and $a_0 \approx 0.65$ fm the diffusivity of the nuclear potential. This expression is expected to provide estimates within a factor of two or so of more accurate treatments. From the deformation length $d_3 = 0.75$ fm associated with the lowest octupole vibration in ^{208}Pb, the formula predicts an excitation cross section of 8 mb for any strongly interacting energetic projectile. This is roughly twice the actual cross section measured in the inelastic proton scattering reaction shown in Fig. 1.12.

7

Compressional modes

In this chapter we study radial oscillations of spherical systems. The oscillatory expansion and contraction of a nucleus is called the breathing mode. As mentioned in Chap. 1, it was identified by the angular distribution of inelastic scattering by nuclear projectiles. It is important to understand this mode in order to obtain information on the compressibility of nuclear matter. We will also discuss the corresponding oscillation for electrons in metal clusters. The radial oscillation behaves as an ordinary plasmon, which classically has a frequency $\sqrt{3}$ higher than the Mie resonance. This mode does not couple strongly to photons, but it may be possible to observe it using inelastic electron scattering.

We begin with some definitions. The compressibility coefficient k is defined as the following derivative of the pressure P with respect to volume or density,

$$k = V\frac{dP}{dV} = \rho\frac{dP}{d\rho} = \rho\frac{d}{d\rho}\rho^2\frac{d(E/A)}{d\rho}. \tag{7.1}$$

In the last equality, we have used the definition of pressure as the derivative of the energy of the system E with respect to volume,

$$P = \frac{dE}{dV} = \rho^2\frac{d(E/A)}{d\rho}$$

Here E/A is the energy per particle of the system. These formulas apply to the ground state compressibility as they stand. For excited systems, to be discussed in Chap. 10, the derivatives are to be interpreted as partial derivatives with respect to density, holding the entropy of the system fixed.

In nuclear physics, the compressibility is usually expressed in

127

terms of the **nuclear compression modulus**, defined by

$$K = \frac{9k}{\rho} \ .$$

Note that the nuclear compression modulus has dimensions of energy; pressure and the ordinary compressibility coefficient both have dimensions of energy per unit volume.

7.1　Nuclear breathing mode: classical

We first present the idealized theory of the breathing mode of a uniform sphere. For radial vibrations of a liquid or solid sphere, the displacement field is given by the gradient of a spherical Bessel function*, $j_0(x)$. The argument of the Bessel function is $x = \omega r / v$ where ω is the frequency and v is the sound velocity. For a liquid there is only one sound velocity, related to the compressibility by $v^2 = k/m\rho$. In the case of an elastic sphere, the longitudinal sound velocity is required, which is related to the Lamé coefficients by the equation, $v^2 = (\lambda + 2\mu)/m\rho$, (cf. eq. (2.30)). The motion satisfies the boundary condition at the surface $\nabla \cdot \vec{u}_0 = 0$ in the case of the liquid sphere and a more complicated condition in the case of the elastic sphere[†].

Let us examine the liquid case in more detail. The boundary condition requires $(d/dx)j_0(x) = 0$ at the surface. This is satisfied for $x = n\pi$, implying the following relation between the lowest frequency and the radius of the sphere R:

$$\omega = \frac{\pi v}{R} \ .$$

Expressing this in terms of the compressibility coefficient k, we obtain the formula

$$\omega = \sqrt{\frac{\pi^2 k}{m\rho R^2}} = \sqrt{\frac{\pi^2 K}{9mR^2}} \ . \tag{7.2}$$

Experimentally, the breathing mode is at 13.9 MeV in the nucleus ^{208}Pb. Taking $R \approx 7$ fm in the above formula, the equation is satisfied for a nuclear compression modulus of $K \approx 210$ MeV.

---

* The Bessel function arises as a solution to the wave equation in spherical coordinates.
† See problem 22.3 in Landau and Lifshitz (1970).

As mentioned in the last chapter, the Fermi "liquid" has characteristics of an elastic medium at high frequencies, so it is far from clear that the liquid drop model discussed in the last paragraph is appropriate. This problem can be sidestepped by considering uniform compressional motion which does not have any shear component. Uniform compression is induced by a field that scales all of the coordinates by the same factor. We write this in terms of a scaling factor $a(t)$ (cf. eq. (2.7)) as

$$\vec{u}(r) = \vec{r}\, a(t) \ . \tag{7.3}$$

As we will see later, the physical displacement field is actually closer to this than to the Bessel function with the hydrodynamic surface condition, due to effects of the finite surface thickness. In any case, the inertia associated with the scaling mode is simply

$$I = \int d^3 r m \rho_0 \vec{u}_0 \cdot \vec{u}_0 = \frac{1}{2} mA <r^2> \ . \tag{7.4}$$

The potential energy function is expressed in terms of the compressibility of the medium as

$$\mathscr{V} = \mathscr{V}(\rho_I) + \frac{kA}{2\rho_0^3}(\rho - \rho_0)^2$$

We express ρ in terms of the scaling variable a, $\rho = (a+1)^{-3}\rho_0$, and find for the restoring force constant

$$C = \frac{d^2 \mathscr{V}}{da^2}\bigg|_{a=0} = \frac{9kA}{\rho_0} = KA \ .$$

This is inserted as the numerator in the Rayleigh principle to produce the following formula for the frequency,

$$\omega = \sqrt{\frac{K}{m <r^2>}} \tag{7.5}$$

It may be seen that this frequency is higher than the hydrodynamic frequency, since $<r^2> \approx 3R^2/5$. Taking the example of ^{208}Pb again, and using the measured charge mean square radius, $<r^2> \approx 30.2$ fm^2, the deduced compression modulus is $K \approx 140$ MeV. The large difference between the scaling and the hydrodynamic estimates is a hint that the relation between nuclear matter compressibility and the breathing mode frequency is not as straightforward as one would like.

Both formulas eq. (7.2) and (7.5) ignore effects associated with the finite surface thickness, which might be expected to be a small correction for a nucleus as large as ^{208}Pb. However, this is not at all the case. It is much easier to compress the surface of a nucleus than the interior, according to any simple model of the nuclear energetics. A model is presented in Appendix C that shows how this happens. For the nucleus ^{208}Pb, the effective compression modulus to be used in eq. (7.5), is only 60% of the nuclear matter compression modulus. Thus it is clear that one cannot make a completely empirical extraction of the modulus from the breathing mode frequency, but must rely on more microscopic theory to account for the surface effects.

7.2   Nuclear breathing mode: RPA

In the last section we considered two simplified models for the displacement field, obtaining similar functional forms for the dependence of the breathing mode frequency on the compression modulus, but rather different numerical coefficients. The disagreement may be understood by going to a quantum mechanical calculation of the frequency. A very systematic study of the breathing mode in RPA was made by Blaizot (1980). He examined a variety of mean field models, each of which produced a ground state of infinite nuclear matter with the required density and energy density. The models differed in the nuclear compression modulus for infinite nuclear matter, varying in the range $K = 200 - 350$ MeV.

Blaizot's analysis clarifies the classical discussion presented in the last section. The first question to be answered by the RPA is, how collective is the breathing mode? This depends on the particular mean field model, but for many of them the vibration is very collective. The strength function for one of the more collective models is shown in Fig. 7.1. The vertical lines show the transition strengths associated with the operator r^2 in the nucleus ^{208}Pb. The dashed lines were calculated from the independent particle response, yielding a strength distribution spread out over the energy region 25-35 MeV. The RPA response, on the other hand, shows a single collective mode at a lower energy that has nearly all ($\approx 95\%$) of the total strength. Qualitatively, this is just the behavior described in Chap. 4 for an attractive residual interaction, cf. Fig. 4.1.

We next look at the displacement field associated with the vibration. This may be defined in terms of the the matrix element

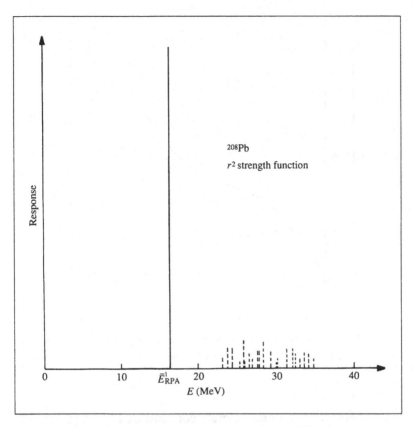

Fig. 7.1. Breathing mode strength function in ^{208}Pb, calculated in RPA as the solid line, and from the independent particle approximation, dashed lines.

of the local current operator between ground and excited state, $\vec{u}(r) \sim < 0|\vec{j}(r)|ex > /\rho\omega$. In fact, it is more convenient to obtain the transition density using the equation of continuity, eq. (3.6), and the relation (3.3) connecting the transition density and the displacement field. The displacement field obtained from the RPA calculations is shown in Fig. 7.2. The dashed-dotted line is the scaling model, eq. (7.3), and the solid line is the result of the RPA calculation. The dotted line is the prediction of the hydrodynamic model, taking the surface at 7 fm. We see that the displacement field follows the scaling model rather closely. This is not surprising in view of the collectivity of the excitation predicted by RPA: according to the sum rule discussion in Chap. 3 (cf. eq. (3.8)), the displacement would have precisely the form eq. (7.3) if the strength

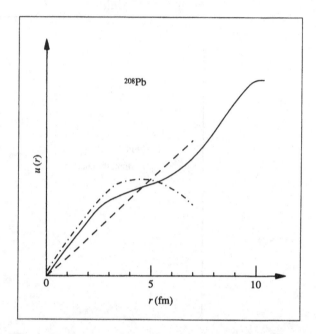

Fig. 7.2. Displacement field $u(r)$ calculated in RPA (solid line), compared with the scaling model (dashed line) and the hydrodynamic model (dot-dash line).

for the operator r^2 were completely concentrated in one state. The modelling as a longitudinal sound vibration is distinctly poorer. The reason is that the nucleus is not uniformly compressible, but is softer in the surface. We will come back to this point in the discussion of the numerical values of the compressibility coefficient.

The transition density associated with the breathing mode has a functional form which may be obtained from eq. (3.3) and (3.6), using the field $\vec{u}_0 = \vec{r}$. One finds

$$\delta\rho \sim \vec{\nabla}\cdot(\delta\vec{j}) \sim \nabla\cdot(\vec{r}\rho_0) = 3\rho_0 + r\frac{d\rho_0}{dr} \ . \tag{7.6}$$

This function is shown in Fig. 7.3 compared to the computed RPA transition density. There are substantial differences in the interior, where the motion induces a maximum compression at the center, as in the hydrodynamic mode. However, in the surface region the computed transition density is indistinguishable from eq. (7.6). This gives support to the scaling model for use in calculating

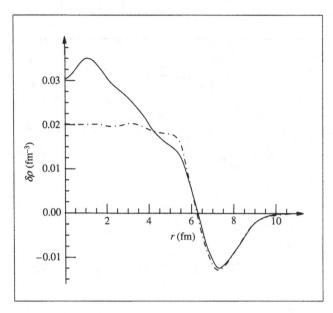

Fig. 7.3. Transition density of the breathing mode of ^{208}Pb. The solid line is the RPA prediction, and the dot-dashed line is the prediction of the scaling model, eq. (7.6).

surface-dominated reactions, such as heavy-ion induced excitation of the breathing mode.

We finally turn to the question of the numerical value of the nuclear compressibility modulus within RPA. If one compares the RPA frequency of the breathing mode with the simple formulas, eq. (7.2) or (7.3), one finds much better agreement with the hydrodynamic formula than with the scaling formula. This seems paradoxical, because the displacement field follows the scaling more closely. The problem is that the nuclear surface is much more compressible than the interior. This both increases the displacement field in the surface and lowers the frequency of the mode from the value given by the bulk formula. The surface plays a very important role even in the largest nuclei one can study. This is discussed in more detail in App. C.

Because of the surface contribution, one would like to understand the A-dependence of the breathing mode as well as possible. If this were known, the extrapolation of the finite systems to infinite nuclear matter could be carried out independently of detailed theory. The systematics of the size dependence of the breathing

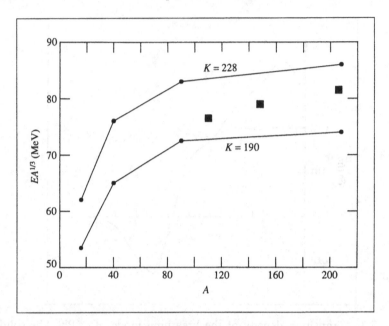

Fig. 7.4. Energy of the breathing mode times $A^{1/3}$ as a function of nuclear mass number A. The data points, representing nuclei Sn,Sm, and Pb, are from Brandenburg, et al., (1987) and Sharma, et al., (1988).

mode are shown in Fig. 7.4. We only display the heaviest nuclei, where the experiments have located the major fraction of the sum rule strength. For comparison, we show the trends of Blaizot's RPA calculations, for two mean field models. These models have nuclear compressibility coefficients of 190 and 228 MeV, and bracket the data. The size dependence in not exactly reproduced, although it is by no means inconsistent in view of experimental difficulties in separating this vibration from other nearby modes.

We conclude with a mention of some more recent work on the breathing mode. In a study by Colò et al. (1992) on ^{208}Pb, a mean-field model was applied that gave rather little collectivity to the mode, although the frequency-compressibility relation followed the Blaizot systematics. The reduced collectivity can be traced back to the particle-hole excitation energies of the mean field model. In the work of Colò et al. this energy is close to $80A^{-1/3}$ MeV, the energy of $2\hbar\omega_0$ excitations in the harmonic oscillator model. The residual interaction is weak, leaving the breathing mode at the same energy.

7.3 Electronic breathing mode

The breathing mode for electrons corresponds quite closely to the ordinary plasmon in metals, which is a longitudinal sound wave. We now estimate the frequency assuming that the displacement field has the scaling form[‡], eq. (7.6). The energy of the compressed electron gas will have two components, a Coulomb interaction and a modified single-electron kinetic energy. The kinetic energy of a scaled wave function varies inversely as the square of the length scale. In the Fermi gas model, this implies an average kinetic contribution to the compressibility given by

$$\frac{d^2}{da^2} \frac{3\hbar^2 k_f^2}{10m(1+a)^2}\bigg|_{a=0} = \frac{9k_f^2}{5m} \; .$$

To get the Coulomb energy, we note that under the compression there will be a uniform interior charge density given by $n = ((1+a)^{-3} - 1)n_0$. This produces an interior electric field $\vec{\mathscr{E}}(r) = \frac{4\pi}{3} n_0((1+a)^{-3} - 1)e\vec{r}$. In the region between the electron sphere and the sphere of positive background charge, the field drops to zero. The total energy may be calculated from the electrostatic field energy, which can be calculated to the needed accuracy neglecting the surface region. This is

$$V = \int \frac{|\mathscr{E}|^2}{8\pi} d^3r = \left(\frac{4\pi}{3}\right)^2 n_0^2 ((1+a)^{-3} - 1)^2 e^2 \frac{R^5}{10}.$$

The restoring force constant for the scaling motion is then given by

$$\frac{d^2 V}{da^2}\bigg|_{a=0} = \frac{8\pi^2}{5} n_0^2 e^2 R^5.$$

Putting this together with the single-particle kinetic energy and the breathing inertia from eq. (7.4), the Rayleigh principle yields for the frequency

$$\omega^2 = \frac{4\pi n_0 e^2}{m} + \frac{3\hbar^2 k_F^2}{m^2 R^2}.$$

[‡] In fact the breathing mode is an exact solution of the classical equations of motion for the radial oscillation of an electron gas.

In the limit of a large sphere, the second term can be dropped and the frequency is exactly the same as a bulk plasmon of infinite wavelength.

This mode does not couple to electromagnetic waves because it is spherically symmetric. In principle, it could be observed by inelastic electron scattering.

8

Spin modes

In previous chapters we discussed collective modes of excitation, emphasizing the connections between classical equations of motion and the quantum physics. Here we want to consider excitation modes whose classical analog is not immediate, the spin and isospin modes. Often in many-fermion systems the particle spins are paired, in which case the single-particle spin operator cannot make excitations. In nuclei, however, the spin-orbital field breaks the spin pairing that would otherwise occur, and the spin operator can induce transitions.

With the nuclear spin-orbit field, the orbital shell l is coupled to total angular momentum $j_> = l + \frac{1}{2}$ and $j_< = l - \frac{1}{2}$, with $j_>$ filled before $j_<$. The σ operator connects the two shells, so particle-hole states of the form $|j_<, j_>^{-1}>$ can be excited in spin-unsaturated nuclei. The residual interaction may be strong in these particle-hole states, producing a rather concentrated strength function as in the giant resonances. However, only the particles of the spin-unsaturated shells participate in the excitation, so it is not as collective as the modes we considered in the previous two chapters.

Similar considerations apply to the charge-exchange operator τ_+. Nuclei with different numbers of neutrons and protons respond to this operator, which typically produces a narrow excited state, the **isobaric analog resonance**. The mode is easily seen with charge exchange reactions, since the excitation is narrow in a region of the spectrum that is rather smooth.

Another mode of nuclear excitation is the Gamow–Teller resonance, excited with the combination of spin and isospin operators, $\sigma\tau_+$. Thus it produces a spin excitation in a neighboring nucleus.

The interest in this mode is largely due to the connection to weak interactions in nuclei. Much of beta decay is mediated by the $\sigma\tau$ operator, and so the characteristics of its strength function are required to understand the beta decay.

If the excitation operator has a spatial dependence in addition to its spin structure, the mode has a classical interpretation as a motion of spin densities, and the reasoning of previous chapters may be applied to understand their properties.

8.1   Isobaric analog resonance

We will apply the method of Chap. 4 to derive a formula for the excitation energy of the isobaric analog resonance. The result, eq. (8.3), is quite simple and may also be derived more elegantly from general considerations of isospin symmetry. However, other modes require the technique of Chap. 4 and this example is instructive. The collectivity of the analog resonance emerges naturally if we use a separable interaction (eq. (4.13). Since the operator exciting the nucleus is τ_+ we use the isospin operator for the interaction field,

$$v(i,j) = \kappa_\tau \sum_{i<j} \vec{t}_i \cdot \vec{t}_j. \tag{8.1}$$

The coefficient κ_τ can be estimated empirically from the isospin dependence of the nuclear optical potential. A commonly used parameterization for the real part of the optical potential is

$$V_{p,n}(r) = \left(-51 \pm 33\frac{N-Z}{A}\right)f(r) \text{ MeV} \tag{8.2}$$

where

$$f(r) = \left(1 + e^{(r-r_0 A^{1/3})/a_0}\right)^{-1}, r_0 = 1.25 \text{ fm, } _0 = 0.65 \text{ fm.}$$

The isospin dependence is contained in the term proportional to $N - Z$, which can be written

$$\pm 33\frac{N-Z}{A} = \frac{33}{A}\tau_z(1)\sum_{i>1}\tau_z(i) \ .$$

Comparing with eq. (8.1), we deduce an interaction strength $\kappa_\tau \approx 33/A$ MeV. It would also be nice to relate the strength of the separable interaction to the underlying nucleon–nucleon interaction. This is a difficult problem because the nucleon–nucleon interaction strongly distorts the wave function when particles are close together. Calculational techniques were developed in the 1960s to deal with the short-range correlations, allowing predictions to be made of the effective interaction to be used with shell-model orbitals*. The most important characteristic of the effective interaction is its volume integral over the relative coordinate between the two particles. One finds for the effective interaction in the isospin channel that the interaction strength has the order of magnitude

$$\int v_\tau(r)d^3r \approx 250 \quad \text{MeV-fm}^3.$$

We express the single-particle potential as an integral over the density and the two-particle interaction as

$$V(r) = \int \rho(r')v(r-r')_\tau d^3r' \approx \rho(r) \int v(r')d^3r' \quad .$$

In the last step, we assumed that the potential has a short range compared to density variations. Applying this to the interior isospin-dependent potential, we find

$$V_\tau \approx \frac{N-Z}{A}\rho_0 \int v_\tau d^3r \quad .$$

where $\rho_0 \approx 0.16$ fm^3 is nuclear matter density. This gives a coefficient of 40 MeV for the symmetry potential, compared with the empirical value of 33 MeV from eq. (8.2).

We now show how the collective analog state arises out of the various particle-hole states that can be made by changing a neutron into a proton. We need only consider states where the quantum numbers of the single-particle orbit are the same for the neutron and proton, that is, configurations of the type $|j_p, j_n^{-1}>$. All these states will be degenerate in energy, and differ in energy from the parent state only by the Coulomb energy and the difference in single-particle potentials for neutrons and protons. The isospin-dependent term in the central potential lowers the energy by $2\kappa(N-Z)/A$,

$$e_p - e_n = V_{coul} + 2\kappa_\tau \frac{(N-Z)}{A} \quad .$$

---

* See Bertsch and Esbensen (1987) for references related to the spin- and isospin-dependent effective interactions.

We still have not included the residual interaction, which connects one particle-hole configuration with another by the matrix element of the operator $\kappa_\tau(\tau_x\tau_x + \tau_y\tau_y) = \kappa_\tau(\tau_+\tau_- + \tau_-\tau_+)$. The dispersion relation associated with the interaction eq. (8.1) is[†]

$$\sum \frac{<jm(p)|\tau_+|jm(n)>^2}{e_p - e_n - \omega} = \frac{1}{\kappa_\tau}$$

where the sum runs over occupied neutron states and empty proton states. The matrix element is simply $<p|\tau_+|n>^2 = 2$, giving a sum $\sum <jmp|\tau_+|jmn>^2 = 2(N - Z)$. It is then easy to see that the dispersion relation has a collective solution with

$$\omega = V_{coul} . \tag{8.3}$$

Furthermore, all particle-hole states (in the m-representation) have equal amplitudes. This is the isobaric analog resonance. As mentioned earlier, eq. (8.3) can also be derived directly without using the explicit particle-hole representation, just invoking the isospin symmetry of the nuclear interaction.

Finally, it should be mentioned that there is a long-standing discrepancy of a few percent between the calculated Coulomb energy in eq. (8.3) and the empirical energy of the isobaric analog state.

8.2   Magnetic modes in nuclei

We now turn to magnetic excitations, for which the dominant operator is $\sigma\tau_z$. The complete magnetic dipole operator is usually expressed in the form

$$\vec{\mu} = \mu_0 \sum_{i=p,n} \left(\mu_{\sigma,i}\vec{\sigma}_i + \delta_{i,p}\vec{l}_i\right) .$$

Here $\mu_0 = e\hbar/2mc$ is the nuclear magneton, the index i sums over neutrons and protons, and $\mu_{\sigma,i} = 2.79$ and -1.91 for protons and neutrons, respectively. It is convenient to express the magnetic dipole operator with isospin operators as

$$\vec{\mu} = \tau_z\mu_0(\mu_v\vec{\sigma} + \vec{l}/2) + \mu_0(\mu_s\vec{\sigma} + \vec{l}/2)$$

[†] The second term in the RPA dispersion relation is missing because the matrix element of the conjugate operator $<p|\tau_-|n>$ vanishes.

The isovector and isoscalar spin moments are $\mu_v = 2.35$ and $\mu_s = 0.44$. Thus the isovector operator is much stronger than the isoscalar.

A separable interaction appropriate for the magnetic excitations may be defined with the spin operators as follows [‡]

$$V_{res} = \frac{1}{2}\kappa_\sigma \sum_{i \neq j} \sigma_i \cdot \sigma_j + \frac{1}{2}\kappa_{\sigma\tau} \sum_{i \neq j} \vec{\sigma}_i \cdot \vec{\sigma}_j \vec{\tau}_i \cdot \vec{\tau}_j.$$

The strength of the interactions κ_σ and $\kappa_{\sigma\tau}$ cannot be determined empirically, as was the case for the isovector interaction. However, we can apply microscopic theory of the residual interaction to predict the strength of the separable interaction, following the further discussion in the last section. The order of magnitude of the volume integrals in the spin channels is $v_{\sigma\tau} \approx 200$ MeV fm^3 and $v_\sigma \approx 0$ MeV fm^3. The deduced strength of the separable interaction in the spin–isospin channel is

$$\kappa_{\sigma\tau} \approx v_{\sigma\tau}\frac{<\rho>}{A} \approx \frac{32}{A} \text{ MeV } .$$

A simple example[§] to illustrate these considerations is the nucleus ^{208}Pb. This is a closed-shell nucleus with two spin-unsaturated shells filled, the $l = 5, j = 11/2$ proton shell and the $l = 6, j = 13/2$ neutron shell. The energy splitting between the two spin-orbit partners can be determined from the spectra of neighboring odd-A nuclei and is nearly the same, $e^p_{j<} - e^p_{j>} = 5.85$ MeV and $e^n_{j<} - e^n_{j>} = 5.57$ MeV. The residual interaction will mix these two excitations quite strongly, making linear combinations that have isoscalar and isovector characters. The isoscalar spin interaction is quite weak, and so the isoscalar state is not shifted much in energy. In fact a state with just the right properties to be this isoscalar spin excitation is found at 5.85 MeV excitation. From the dispersion relation one may evaluate the excitation energy of the isovector state E_v. The expression neglecting the second term in the propagator is

$$E_v \approx e_{j<} - e_{j>} + \frac{16}{3}\kappa_{\sigma\tau}\frac{l(l+1)}{2l+1} . \qquad (8.4)$$

[‡] The reader may wonder why the orbital operator was not also included in the separable interaction. While the nuclear interaction is strongly spin-dependent, its dependence on momentum is much weaker, so the coefficient of l in a separable expansion would be small.

[§] We follow here the discussion of Wambach (1988).

Inserting numbers with average values for the energies and the angular momenta, one finds $E_v \approx 7.6$ MeV. Experimentally, a concentration of transition strength is indeed found at this energy.

Further information about the modes comes from the transition strengths. The theoretical transition strengths for the isoscalar and isovector modes of the same l are given by

$$B(M1, 0 \to 1^+)_{s,v} = \frac{3}{4\pi} \sum_m <0|\vec{\mu}|1^+, m>^2$$

$$= \frac{3\mu_0^2}{4\pi} [4\mu_{s,v} - 1]^2 \frac{l(l+1)}{2l+1} .$$

The numerical result is $B(M1)_s = 0.4\mu_0^2$, and $B(M1)_v = 50.2\mu_0^2$, where we have again taken an average $l \approx 5.5$. The observed mode at 5.85 MeV has somewhat more strength than this ideal isoscalar mode, with $B(M1) = 1.6 \pm 0.5\mu_0^2$. This is understandable in terms of the incomplete degeneracy of the proton and neutron states, allowing some isovector strength in the transition. The isovector transition, which is spread over a number of states between 7 and 8 MeV, has a total strength of only $20\mu_0^2$, less than half the predicted value. This reduction has been much discussed, and there are a number of effects that contribute to it. The predicted transition probabilities can be significantly affected by modifications of the ground state wave function due to the residual interaction. The main effect is to reduce the transition probability associated with the isovector transition by ≈ 20–30%, that is, $B(M1) \approx 40\mu_0^2$. A further reduction of this strength comes about because the mesonic and other nuclear many-body effects, modify the free nucleon g-factors. This quenching will also be seen in the next section.

8.3 Gamow–Teller resonances in nuclei

The operator $\sigma\tau_{\pm}$ also arises in beta decay, inducing the so-called Gamow–Teller transitions. Invariably these transitions are hindered with respect to the single-particle rates predicted from the independent particle shell model. The analogous quenching of low electric dipole transitions is understood in terms of the concentration of $E1$ strength in the giant dipole resonance. The

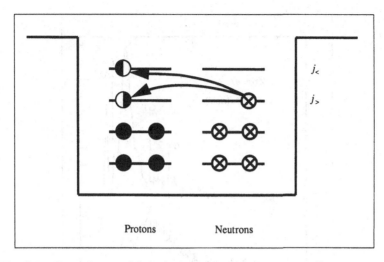

Fig. 8.1. Particle transitions induced by the Gamow–Teller operator.

corresponding resonance which exhausts the Gamow–Teller β-decay strength is known as the Gamow–Teller giant resonance.

The beta decay process does not have enough energy to excite the giant resonance, but it can be produced by the (p, n) reaction at intermediate energies. The incoming proton exchanges charge with the target and becomes a neutron. Experiments also show that spin is preferentially transferred; most of the outgoing neutrons have a spin direction different from that of the incoming proton. Consequently, in the excitation of these resonances the nuclear structure enters through the matrix elements of the charge- and spin-changing operators

$$\rho_{\sigma\tau} = \sum_i \sigma_i \tau_{\pm} \delta(\vec{r} - \vec{r}_i) \ .$$

The action of the $\sigma\tau_-$ operator on a nuclear ground state configuration is shown schematically in Fig. 8.1.

The (p, n) experiments

Fig. 8.2 shows spectra at different angles from (p, n) reactions on ^{208}Pb. The spectrum in the forward direction is dominated by a single peak. As the scattering angle increases, this peak diminishes in size and at angles $\theta = 4\text{-}5^o$ is replaced by another broader peak at lower neutron energy, i.e. higher excitation energy. As one goes

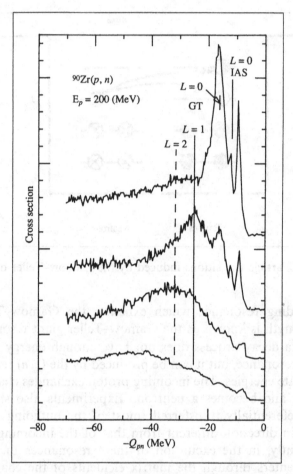

Fig. 8.2. Spectra of neutrons from the (p, n) reaction on ^{90}Zr at 200 MeV bombarding energy, from Gaarde (1981). The scattering angles for the different curves are 0.5°, 4.5°, 9.5°, and 14.5°, from top to bottom.

to larger angles, the peak moves to progressively higher excitation energy and becomes even broader.

The peak in the zero degree spectrum is the Gamow–Teller giant resonance. For the nucleus ^{90}Zr with 40 protons and 50 neutrons, the excitation involves the 10 excess neutrons, which can coherently transform into protons and change their spin direction. On the microscopic level, the excess neutrons in the $g_{9/2}$ orbital can be transformed into protons and put in the $g_{9/2}$ or $g_{7/2}$ orbitals. The proton particle and the neutron hole couple to $J^{\pi} = 1^{+}$. The fact that the peak disappears away from the forward scattering direction shows that the transition density is very smooth. If there were strong spatial variations in the transition density, the scattering process would favor non-zero momentum transfer and finite scattering angles. One can confidently infer from the angular distribution that the mode carries no orbital angular momentum.

The angular distribution shows that the peak at higher energy is associated with transitions having nonvanishing orbital angular momentum. For example, the peak at 4.5° spectrum in Fig. 8.2 is interpreted as an envelope of the giant resonances with orbital angular momentum $L = 1$ and total angular momentum 0, 1 and 2. In a microscopic description, the excitation could, for example, be a transition from a $g_{9/2}$ neutron orbital to an $h_{11/2}$ proton orbital. This is a transition from one shell into the next. In ^{90}Zr, the energy difference between shells is 6 MeV higher than for the $L = 0$ resonance.

The (p, n) spectra show very small cross-sections for transitions to states where spin transfer is not allowed, for example, $0^+ \rightarrow 0^+$ transitions. The spectra given in Fig. 8.2 are therefore totally dominated by spin transfer processes.

We implied above that the forward-angle (p, n) reaction measures Gamow–Teller strength. This has been demonstrated by measuring the (p, n) cross-section for a number of transitions for which the β-decay has been studied (Goodman et al. (1980)). From the lifetime and energy of the β-decay the strength can be deduced, and a comparison with the (p, n) zero degree cross-section shows a close proportionality.

Sum rule

The Gamow–Teller transition strengths satisfy an important sum rule. It relates the difference between the strengths for β_- and β_+ transitions from some initial state:

$$S_{\beta^-} - S_{\beta^+} = 6(N - Z). \qquad (8.5)$$

where the strength S is defined

$$S_{\beta\pm} = \sum_f <i|\vec{\sigma}\tau_\pm|f>^2 .$$

The sum is made over all final states in both the β_- and β_+ channels. The sum rule is derived from the commutator relation $[\tau_-\sigma_i, \tau_+\sigma_i] = 2\tau_z$, following the same procedure as in Sect. 3.5.

Two strengths are required to test the sum rule, $S_{\beta+}$ and $S_{\beta-}$, but only the $S_{\beta+}$ can be measured by the (p, n) reaction. However, we know that $S_{\beta+}$ is a positive number, and consequently $6(N-Z)$ is a lower limit for $S_{\beta-}$. One can actually make sharper use of the sum rule, however. In nuclei with large neutron excess like ^{208}Pb, $S_{\beta+}$ is small because the final neutron states in which the β-decay protons could decay are already occupied, and are blocked by the Pauli principle. In effect, the sum rule needs to be satisfied for $S_{\beta-}$ alone.

The detailed analysis of the (p, n) spectra shows that only between 50% and 65% of the β^- strength deduced from the sum rule is found in the expected energy region of the spectrum. Two mechanisms to explain the missing strength have been discussed. The first is the coupling of the Gamow–Teller resonance to more complicated configurations. This removes strength from the low-lying region and spreads it thinly over a large energy region (Bertsch and Hamamoto (1982)). In the ^{90}Zr spectrum the missing strength would then be in the long tail towards higher excitation energy, i.e. lower outgoing neutron energy. Although there is no experimental proof of Gamow–Teller strength in this tail region, the possibility cannot be ruled out. The second mechanism is a possible role of the internal structure of the nucleons. The simplest excitation of the nucleon is the Δ-resonance with a mass of 1232 MeV, which in a quark model is a spin–isospin excitation where a d-quark is transformed into a u-quark. One would therefore expect that in a process involving spin–isospin excitations the Δ-resonance would also be excited. According to some estimates, this mechanism can remove about $\approx 30\%$ of the strength (Bohr and Mottelson (1981)). To date, no direct confirmation of the workings of this mechanism has been observed.

8.4   Spin in metal clusters

We conclude this chapter with a few remarks on manifestations of the electron spin in metal clusters. In principle the spin is observable using the Stern–Gerlach technique, deflecting the clusters in an inhomogeneous magnetic field according to their spin orientation. In alkali metal clusters, no spin has been detected for clusters larger than three atoms. Since odd-N clusters certainly have an unpaired electron spin, this shows that the spin is strongly coupled to other degrees of freedom, so that the expectation value of the spin in any direction is very small. In contrast, clusters of atoms such as iron that are magnetic in bulk show strong deflection in the Stern–Gerlach measurement. However, unlike the classical measurement in which both positive and negative deflections were observed, the magnetic clusters are always attracted into the region of the strong field (de Heer et al. (1990)). Again, this shows that the spin is strongly coupled to other degrees of freedom, but details can only be unraveled by studying the dependence on excitation, angular momentum, and other variables.

The study of magnetic excitations in clusters requires a probe that couples to the electron spin. Inelastic electron scattering is dominated by the Coulomb interaction, but the exchange term is present for parallel electron spins, and thus introduces a spin dependence. The exchange interaction is strongly dependent on the energy of the probing electron, becoming small for high-energy electrons. This gives a way in principle to recognized magnetic excitations in inelastic electron scattering; their peaks in the energy loss spectrum would be more prominent for lower energy electrons than for high energy electrons.

Magnetic states of orbital nature in deformed metal clusters have been discussed by Lipparini and Stringari (1989). Collective spin excitations have been suggested by Serra et al. (1993).

9
Line broadening and the decay of oscillations

There are many physical processes that damp oscillations, producing a finite width to the spectral lines. These will be cataloged and described in some detail in this chapter. The processes may be divided into two broad categories, depending on whether the energy of the excitation escapes from the system, or whether it is merely redistributed into other degrees of freedom within the system. In the first category we have the natural decay processes such as photon emission or particle emission. Widths due to photon emission are completely negligible if there are any competing damping mechanisms, but particle emission widths can be important in small systems. The second category includes a number of diverse mechanisms for spreading the strength function of collective oscillation, depending on the particular degrees of freedom that are involved. First to be mentioned are the single-particle degrees of freedom. These damp the collective motion, if their energy spectrum is dense near the collective excitation energy. The single-particle damping is the most important damping mechanism for charge oscillations in classical hot plasmas, and is known as Landau damping. Another damping mechanism corresponds to collisions between particles within the system. Quantum mechanically, this mechanism appears when configurations having a more complex character mix into the simple single-particle single-hole configurations that characterize the collective excitation. As a rule these collisions are more effective when they take place at the surface of the system. This is because the density is lower and the Pauli principle is less effective in blocking final states. Also collective oscillations of the surface may be excited, effectively enhancing

148

the collisional damping. In the electronic excitations of molecules and atomic clusters, coupling to the motion of the atomic centers provides yet another additional damping mechanism.

The important role played by the surface in the purity of a collective vibration is particularly simple to see in the case of nonspherical systems. Under such conditions, the vibrational frequency depends on the orientation of the oscillation with respect to the axes of the system. The observed oscillation will be a superposition of the normal modes at different frequencies. When the system is heated, thermal fluctuations produce an ensemble of shapes and the line appears broadened. This broadening is a property of the ensemble, not necessarily of an individual cluster, analogous to "inhomogeneous line-broadening" in macroscopic systems.

In our discussion of the damping mechanisms in this chapter, we shall make frequent use of Fermi's Golden Rule, the perturbation formula for the width of a state. The formula expresses the decay rate of a state τ and its spectral width Γ as

$$\Gamma = \hbar/\tau = 2\pi \overline{<i|v|f>^2} dn_f/dE. \tag{9.1}$$

Here v is the residual interaction between states, with i the collective excitation and f any other state, $\overline{<i|v|f>^2}$ is the average squared matrix element, and dn_f/dE is the spectral density of states. This formula is valid when there is a high density of states f that mix with the collective excitation, and the variation of the quantities with energy is small. The predicted line shape in this case follows the Breit–Wigner form, eq. (2.36),

$$S(E) \sim \frac{1}{(E - E_r)^2 + \Gamma^2/4} \ .$$

We shall also meet situations in which the density of final states in not high enough to satisfy the conditions of the last paragraph. In many of these cases it is useful to consider the variance of the strength distribution in the presence of the coupling to the other degrees of freedom. This is given by

$$\sigma^2 = \sum_f <i|v|f>^2 \ .$$

When there are many independent couplings with finite dispersions, the overall strength distribution approaches a Gaussian shape,

$$S(E) \sim e^{-(E-E_r)^2/2\sigma^2} \, ,$$

and the associated full width at half maximum is given by

$$\Gamma = \sqrt{8\log_e 2} \, \sigma \, . \tag{9.2}$$

9.1 Particle escape width

If the frequency of oscillation is higher than the binding energy of a particle, particle emission will produce a finite lifetime and damp the oscillation. This is the situation in the nuclear giant resonances and in the atomic "giant resonance" depicted in Fig. 1.3. Qualitatively, there are three regimes of behavior of the escape width.

When there is essentially no barrier to the escape of particles from the system, the excitation is so broad that the concept of a collective oscillation is hardly useful. An example is the dipole excitation of the small nucleus ^{4}He. In the shell model, the nucleons are excited from the s-wave ground state to a p-wave state in the continuum. The energy is so high that the centrifugal barrier of the p-wave plays no role and the nucleons are essentially free. Under these conditions the width of an excitation can be estimated from a simple classical argument. We assume that the energy of the oscillation is in kinetic energy of an excited particle, which moves with a velocity v. We place the particle randomly in a spherical volume of radius R, and calculate the rate at which it crosses the surface. This is the product of the density of the particle inside the surface area, and the average projection of its velocity perpendicular to the surface. The estimate for the rate W is

$$W = \frac{\rho A}{4\pi} \int_0^{2\pi} d\phi \int_0^1 d\cos\theta \, v \cos\theta = \frac{3v}{2R}.$$

Converting this rate to an energy width, we obtain the following single-particle estimate,

$$\Gamma^\uparrow = \frac{3\hbar v}{2R} \, . \tag{9.3}$$

In the example of the dipole excitation of ^{4}He, we can make a numerical estimate taking the average energy of the neutron to be

5 MeV and a radius of the ^{4}He to be $R = 2$ fm. Then the above equation gives a width

$$\Gamma^\uparrow \approx (3 \times 200\text{MeV fm/c})(0.1c)/(2 \times 2\text{fm}) = 15\text{MeV}.$$

From this estimate we see that the excitation is very broad, and in fact hardly can be described as a resonance at all.

It is more usual that a few particles are weakly bound and can be emitted, but most particles would remain bound after absorbing the energy of the collective excitation. Then the peak can be rather narrow, with the width governed by the rate at which the collective field puts energy from the closed channels into the open channels. This is the situation for the giant dipole resonance of ^{16}O. The resonance is peaked at an energy of 23 MeV, which is 7 MeV above the neutron emission threshold. However, only the particles from the $p_{1/2}$ shell are in the continuum at this energy. We may estimate the rate at which the closed channels feed into the open s-particle $p_{1/2}$-hole channel as follows. We apply Fermi's Golden Rule to the open channels, taking the transition potential δV as the perturbation. Eq. (9.1) takes the form

$$\Gamma = 2\pi \overline{<\phi_b|\delta V|\phi_c>}^2 \frac{dn_c}{dE}$$

where ϕ_b is the wave function of the bound particle that is excited into the continuum, ϕ_c is the continuum wave function, and dn_c/dE is the density of states in the continuum.

To make a rough estimate, we will first scale ϕ_c to a corresponding wave function that is normalized within the nucleus. The matrix elements for normalized wave functions will then be related to the total interaction energy of the oscillation, which is a known physical quantity. We assume that ϕ_c is an s-wave function and neglect potential interactions. Then it is proportional to a sine function; including the continuum normalization factor it is

$$\phi_c \sqrt{\frac{dn_c}{dE}} = \sqrt{\frac{2m}{\hbar^2 \pi k}} \sin kr$$

where k is the reduced wavenumber of the particle in the continuum. Of course, the particle wave function should be an eigenfunction of the potential well; the effect of the distortion inside the system roughly changes k, the external wave number, to K, the wave number inside the nucleus. We compare our continuum wave function with a simple idealized bound state wave function,

$$\phi'_b = \sqrt{\frac{2}{R}} \sin k'r \ .$$

To determine the normalization factor $\sqrt{2/R}$, we have assumed that the wave function vanishes at R, the nuclear surface radius. We then obtain the following estimate for the continuum matrix element in terms of corresponding bound state matrix elements,

$$<\phi_b|\delta V|\phi_c> \sqrt{\frac{dn_c}{dE}} \approx \sqrt{\frac{mR}{\hbar^2 \pi K}} <\phi_b|\delta V|\phi'_b> \ .$$

The bound state matrix elements may be estimated from the collective frequency shift of the vibration as

$$\Delta E = <\phi_b|\delta V|\phi'_b> \sqrt{N}$$

where N is the number of bb' configurations that participate in the collective oscillation (cf. eq. (4.16). We thus obtain for our estimate of the width

$$\Gamma = \frac{2mR(\Delta E)^2}{\hbar^2 K} \frac{N_{open}}{N} \tag{9.4}$$

where N_{open} is the number of open channels.

The giant dipole resonance in a light nucleus has a high excitation energy relative to the binding energy of the particles at the Fermi surface, and many of the channels are open. This gives a relatively large escape width for these nuclei. In our example of ^{16}O, the parameters are: $\Delta E \approx 8$ MeV, $R \approx 3$ fm, $N_{open}/N \approx 1/3$, and $K \approx 1.5$ fm^{-1}. The resulting width is $\Gamma \approx 2$ MeV. This computation is too crude for anything but an order-of-magnitude estimate; obviously the details of the continuum wave function in the potential field will be important, and each open channel has a separate contribution to the width. Empirically, the dipole resonance has structure associated with the closed channels on the scale of 3–5 MeV, together with a broad high energy tail associated with the open channels. This may be seen in Fig. 9.1. In very heavy nuclei, the escape width becomes small because the collective resonance energies decrease, leaving very few open channels.

A number of methods are available for a more quantitative description of the particle continuum. One of the simplest methods is to calculate the RPA response function in coordinate space, as described in Chap. 4. The RPA theory describes single-particle

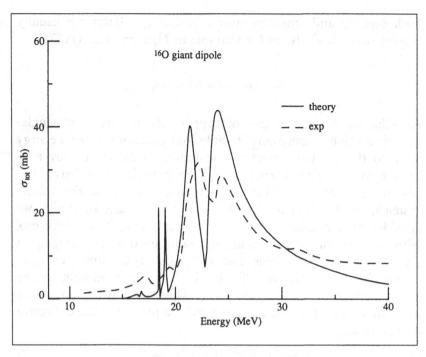

Fig. 9.1. Giant dipole resonance in ^{16}O. Dashed line: experimental; solid line, continuum RPA theory (Shlomo and Bertsch (1975)).

motion in a time-varying potential, and in principle the theory is the same, whether or not the particles are bound. If the Green's function approach is used to solve the equations, the only difference between bound and unbound is the boundary condition on the Green's function. This method has been used by several authors to make theoretical predictions of line shapes in nuclear excitations and in electronic excitations in alkali metal clusters. See App. E and references there for more details. An example of the continuum RPA is shown in Fig. 9.1, the predicted strength function for the ^{16}O giant dipole resonance.

For many purposes the coordinate-space response function method is too restrictive, and the RPA theory is calculated in a configurational representation. This requires a discrete basis of single-particle states, so states in the continuum must be discretized in some way. It is still possible to calculate the full continuum response, defining a single-particle Hamiltonian with both discrete and true continuum states. The latter are usually treated perturbatively, as for example in Nguyen et al. (1987).

9.2 Landau damping

A collective oscillation can disappear into uncorrelated single-particle motion if many particle-hole configurations have an energy close to that of the collective oscillation. This degeneracy may arise from the kinetic energies of the particles in a Fermi gas description, which is the Landau damping mechanism. For infinite systems, the criterion for Landau damping is quite simple. If we consider an excitation with some momentum k, it can only mix with particle-hole states having the same total momentum. In a Fermi gas, states of momentum k can be made having energies up to $e(k_f + k) - e(k_f) = \hbar^2(2k_f k + k^2)/2m$. The minimum energy is zero, providing k is not too large. Thus, for not too large k in a Fermi system, Landau damping will be present if the collective energy satisfies

$$\hbar\omega_c \leq \hbar^2(2k_f k + k^2)/2m \ . \tag{9.5}$$

In finite systems the situation is much more complicated because the boundaries produce shell effects in the density of particle-hole states. These shell effects can nullify the Landau damping for the most collective oscillations. A very crude picture, applicable to the smallest systems, can be made using the harmonic oscillator model for the single-particle energies. Here the particle-hole energies are all concentrated at multiples of the oscillator frequency. For example, in the case of the nuclear giant dipole, the resonance energy, eq. (5.11), is approximately twice the oscillator frequency, eq. (A.2). This puts the mode in a gap between energies of unperturbed particle-hole states of the same quantum numbers, which occur at odd multiples of $\hbar\omega_0$. Thus there is relatively little Landau damping for the nuclear giant dipole. Of course, the oscillator picture is not perfect. There are states within the gap because the spin-orbit field in nuclei changes the single-particle spectrum substantially away from the oscillator form. The structure in the strength function of Fig. 9.1 is due to this effect. Another example of an RPA strength function is shown in Fig. 9.2, displaying the predicted dipole response for the heavy nucleus ^{208}Pb. The main single-particle transitions carrying dipole strength have unper-

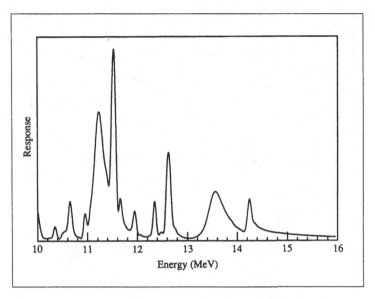

Fig. 9.2. Dipole response in RPA for the nucleus ^{208}Pb. The calculation was done with the program RPA3, described in App. E.

turbed energies near $\hbar\omega_0 \approx 7$ MeV, but there are some transitions such as $i_{13/2} \rightarrow j_{13/2}$ that lie in the region 12–14 MeV, where the giant dipole resonates*.

The situation for the Mie resonance in metal clusters is rather different. Here the ratio of resonance energy to oscillator energy may be estimated from eq. (5.6) and (A.3) as $(\hbar\omega_M)/(\hbar\omega_0) \approx 0.58\sqrt{r_s}N^{1/3}$ where N is the number of active electrons, and r_s is the Wigner–Seitz radius. Landau damping will be reduced when the left hand side is a small even integer, for the same arguments as given for the nuclear giant dipole. This is illustrated schematically in Fig. 9.3. Taking the empirical Mie resonance energy in sodium clusters, the gap at $2\hbar\omega_0$ occurs near $N = 14$. In the spherical cluster Na$_8$ the resonance is a single state, and in in the $N = 20$ cluster it is split into 2. In the charged potassium cluster K_{21}^+, the resonance is a single state, as may be seen from Fig. 9.4. On the other hand, for $N = 40$ the resonance energy is at $2.8\hbar\omega_0$, where the level density of particle-hole state should be high. Fig. 9.5

---

* The single-particle Hamiltonian in this calculation used a Woods–Saxon potential well. A nonlocal Hartree–Fock potential would have a higher density of particle-hole states in the giant dipole region.

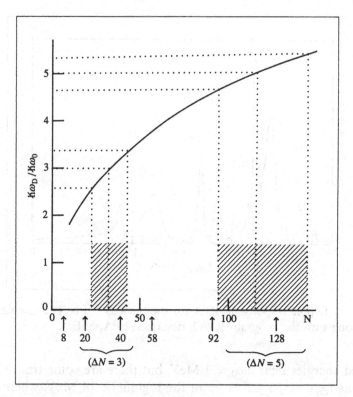

Fig. 9.3. Particle-hole density of states for alkali metal clusters. Shaded regions have a high level density in the vicinity of the Mie resonance.

shows the prediction of RPA theory in this cluster, and it may be seen that the resonance here is split into many components.

When the size of a metal cluster becomes so large that the Mie frequency is high compared to the basic shell frequency, the theory simplifies considerably. The Landau damping will then behave in a smooth way, because many shells will participate in the damping and the shell fluctuations in the level density will average out. The theory for the damping in this limit was given by Kawabata and Kubo (1966); certain approximations in the original formula were improved upon by Barma and Subrahmanyam (1989) and by Yannouleas and Broglia (1992).

We now derive this damping formula, proceeding along the lines developed earlier in this chapter. Using eq. (9.1), we write

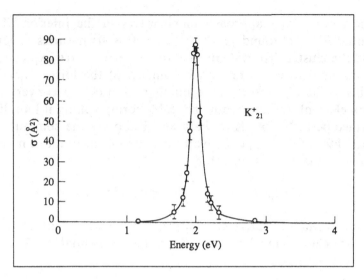

Fig. 9.4. Mie resonance in the charged potassium cluster K_{21}^+, from Brechignac et al. (1989).

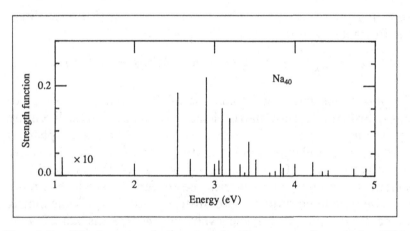

Fig. 9.5. Dipole strength function in Na_{40}, according to the RPA calculation of Yannouleas and Broglia (1991).

the width as

$$\Gamma_L^\downarrow = 2\pi \sum_{p,h} <p|\delta V|h>^2 \, \delta(\hbar\omega_M - e_p + e_h) \qquad (9.6)$$

where p, h label particle and hole states. Because the system is large, we may neglect the shell effects in the state $|p,h>$ and treat them as the eigenstates of a semi-infinite potential which vanish in the exterior and approach plane waves in the interior. The potential δV was found in Chap. 5, eq. (5.8-9); it varies linearly inside the cluster. To evaluate the above perturbative expression, we thus need to evaluate matrix elements of the form $<i|z|j>$. At this point one resorts to a number of tricks to convert the matrix element into a more manageable form, making it look like a surface perturbation. Kawabata and Kubo use the commutator identity $\hbar\nabla_z/m = [H, z]$ to convert the matrix element into a matrix element of the current operator,

$$<i|z|j>= \frac{\hbar}{m(e_i - e_j)} <i|\nabla_z|j> \quad .$$

We transform the matrix element once more with another commutator relation involving the mean field potential V, $\nabla_z V = -[H_0, \nabla_z]$. This gives

$$<i|z|j>= \frac{\hbar^2}{m(e_i - e_j)^2} <i|\nabla_z V|j> \quad .$$

Inserting this in eq. (9.1), with the δV from eq. (5.8), we obtain the following expression for the damping width,

$$\Gamma_L^\downarrow = \frac{\pi}{2N\omega_M m} \sum_{p,h} <p|\nabla_z V|h>^2 \delta(\hbar\omega_M - e_p + e_h) \quad .$$

To obtain this result, we made use of the δ-function to replace e_p-e_h by $\hbar\omega_M$. The last matrix element is evaluated in the literature by writing out the single-particle wave functions of a spherical potential well and explicitly integrating with the surface derivative potential, dV/dr. This requires knowledge of the properties of spherical harmonics and spherical Bessel functions, which is more than we wish to go through here. Instead, let us simplify the matrix element by setting $j = i$ and evaluating dV/dr instead of $\nabla_z V$. This matrix element is trivial to evaluate using the Feynmann–Hellman theorem, $dE_i/dr =< i|dH_0/dr|i >$. The energy of the state in a spherical well is just the kinetic energy, and for a fixed nodal structure, that depends on the radius of the well as r^{-2}. Thus the matrix element is

$$<i|dV/dr|i>= \frac{2e_i}{R} \quad .$$

Of course, we really want the off-diagonal matrix elements of the operator $\nabla_z V = dV/dr \cos\theta$. Let us examine the squared off-diagonal matrix element, summed over final angular momentum states,

$$\sum_{l'} <l|\frac{dV}{dr}\cos\theta|l'>^2 \ .$$

The angular integral is

$$\int d\Omega d\Omega' \, Y_l^*(\Omega) \, Y_l(\Omega') \cos\theta \cos\theta' \sum_{l'm'} Y_{l'm'}^*(\Omega') \, Y_{l'm'}(\Omega)$$

$$= \int d\Omega |Y_l(\Omega)|^2 \cos^2\theta = \frac{1}{3} \ ,$$

where to complete the last step we averaged over m-states of the initial states. The radial matrix element we take the same as the diagonal matrix element, to obtain

$$\sum_j <i|\frac{dV}{dr}\cos\theta|j>^2 = \frac{4e_i^2}{3R^2} \ .$$

The remaining tasks are to sum over final states with different radial nodes and to sum over hole states. The sum over hole states is straightforward; all occupied orbitals can be excited that satisfy $e_h + \hbar\omega_M > e_f$. For the limiting case $\hbar\omega_M << e_f$, this sum is

$$\sum_h = \frac{3N\hbar\omega_M}{2e_f} \ .$$

The density of particle states of fixed angular nodal structure is the only remaining unknown quantity. This becomes, at the Fermi surface,

$$\sum_p \delta(e_p - e_h - \hbar\omega_M) = \frac{R}{\pi}\frac{dk}{de} = \frac{Rm}{\hbar^2 k_f \pi}.$$

Putting all these ingredients together in eq. (9.6), we end up with a simple formula for the width,

$$\Gamma = \frac{\hbar^2 k_f}{mR} \ . \tag{9.7}$$

Thus the damping rate is given by the traversal time of the particles across the cluster, traveling at the Fermi velocity $v_f = \hbar k_f/m$. The formula has the same dimensional structure as the single-particle escape width, eq. (9.3). This is quite natural because in both cases the energy of the collective motion is being transformed by the surface. The very different physics responsible for the two damping

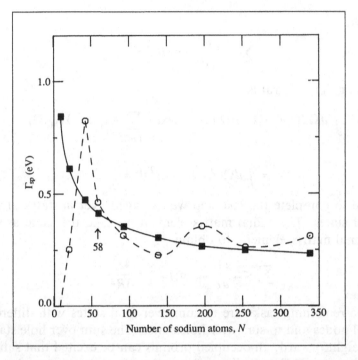

Fig. 9.6. Width of Mie plasmon in Na clusters. The solid line is the $1/R$ law, similar to eq. (9.7), and the open circles were extracted from RPA calculations of the dipole strength function(Yannouleas, Vigezzi and Broglia (1993)).

processes is reflected only by the numerical factors. As mentioned earlier, more careful evaluation of the various approximations gives slightly different numerical coefficients. A comparison of the $1/R$ law with microscopic RPA calculations is shown in Fig. 9.6. In detail the RPA results will depend on the specific Hamiltonian model and on how a width Γ is extracted from the RPA strength function, but one can see that these results fluctuate in the vicinity of the expected values. The fluctuations are presumably the shell effects discussed above.

The inverse relation between the peak width and the size of the cluster is seen very well in the systematics of silver clusters, Fig. 9.7 from Genzal et al. (1975). However, this data is from clusters embedded in a matrix and the coefficient depends on the matrix, i.e. on the conditions at the surface. The dependence on the environment is unexpected; the derivation did not make use

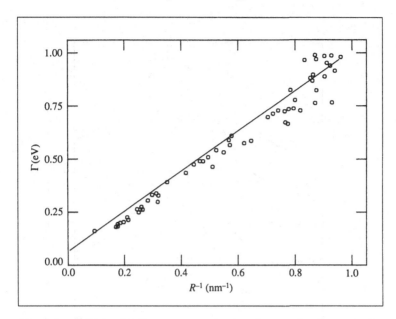

Fig. 9.7. Width of Mie resonance in Ag clusters in glass matrix.

of the detailed properties of the surface, but only that it was a boundary for the electrons.

9.3 Deformation effects

In both nuclei and atomic clusters deformations have a marked effect on the dipole oscillations. We have earlier derived the dependence of the frequency on the axis length for an ellipsoidal shape. In this section we want to describe the effect of shape fluctuations on the oscillation. Obviously, if the resonance frequency depends on shape, there will be a distribution of frequencies if the shape can change. Empirical evidence for this spreading may be seen in the evolution of $\Gamma^{\downarrow}$ in the neodinium isotopes as a function of mass number (cf. Fig. 9.8). The only nucleus of this series that is deformed in the ground state is ^{150}Nd, but the lighter ones become progressively softer with respect to quadrupole fluctuations of the surface. This is evidenced by the progressively lower energy of the lowest 2^{+} state. In the series of isotopes from ^{142}Nd to ^{148}Nd, the giant dipole increases in width from 4.4 MeV to 7.2 MeV.

A solvable model for the effect of shape fluctuations can be

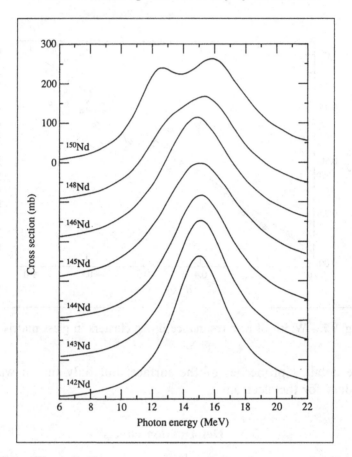

Fig. 9.8.   Giant dipole resonance in neodinium isotopes, after Berman and Fultz (1975).

made considering the surface vibrations to be harmonic modes of some frequency ω_v and assuming that the dipole vibration is a single state that couples linearly to the deformation. Then the spectrum consists of a sequence of peaks at frequencies $\omega_D + n\omega_v$. The strength of the individual peaks is given by the Poisson distribution[†] The important quantity in determining the character of the spectrum is the ratio of the frequency shift of the dipole state at the r.m.s. ground state deformation to the frequency of the surface mode.

[†] Mahan (1981) gives an exposition of the theory for linear coupling to the harmonic oscillator.

In the adiabatic limit, the surface mode frequency is very small compared to the dipole state frequency shifts. Then the Poisson distribution becomes close to a Gaussian, and we may characterize it by the second moment of the strength function. This moment is given by an obvious formula, the expectation of the dipole frequency squared evaluated in the ground state wave function of the shape degree of freedom. Generalizing this to the five independent degrees of freedom of quadrupole shape vibrations, we have

$$\sigma_D^2 = \hbar^2 \int \phi_0^2(\beta)(\omega_K(\beta, \gamma) - \omega_D)^2 d\tau$$

where $\phi_0(\beta)$ is a wave function to describe the surface fluctuations, $\omega_K(\beta, \gamma)$ is the dipole frequency of the K-mode for a deformed nucleus with shape parameters β and γ, and $d\tau$ is the volume element for integration in the β, γ plane (Bohr and Mottelson (1975)). We evaluate the integral over the oscillator ground state wave function ϕ_0 and obtain

$$\sigma_D \equiv <(\Delta \hbar \omega_D)^2>^{1/2} = \sqrt{\frac{5}{8\pi}} \beta_0 \hbar \omega_D \qquad (9.8)$$

where β_0 is the root mean square deformation associated with the zero-point motion of the ground state. This is just the same as we would obtain for the splitting due to a static deformation of the same amount. Assuming the distribution to be Gaussian, we may associate a width with the variance according to eq. (9.2). For typical cases, $\beta \approx 0.2$, which for a nucleus of mass $A = 150$ leads to a damping width $\Gamma^\downarrow \approx 3$ MeV. The adiabatic limit is clearly applicable here, since the excitation energy of the quadrupole vibrations in soft nuclei is only about 0.5 MeV. What is not so clear is how to combine this width with the contributions from other sources. If all the other degrees of freedom behave adiabatically, the widths should be added in quadrature and then the increase in ^{148}Nd (from 4.4 MeV in ^{142}Nd) would only be 1 MeV. On the other hand, if the widths represent true decays channels, they should be added directly. This would give 7.4 MeV, close to the observed width of ^{148}Nd.

The situation in metal clusters is rather similar (cf. Pacheco, et al. (1991)). Here, the zero-point motion in the shape degrees of freedom is described by the normal modes of vibration. The amplitude of the motion is small, however. To give some idea of the magnitudes involved, let us consider the cluster Na_{20}. The r.m.s. fluctuation in ϵ is given by

$$<\epsilon^2>^{1/2} = \sqrt{\frac{\hbar}{2I\omega_v}}$$

where I is the inertia from eq. (2.26). In Chap. 2, following eq. (2.31), we also estimated the vibrational frequency and energy, $\hbar\omega \approx 3.6$ meV.

Inserting the numerical values in the above equation, we find for the dispersion in ϵ, $< \epsilon^2 >^{1/2} \approx 0.01$. The shift in the plasmon frequency for this deformation, from eq. (5.14), is $< (\omega - \omega_D)^2 >^{1/2} \approx 35$ meV. This width is larger than the quantum of excitation energy, implying that the strength will be spread over several vibrational states. Empirically, the observed width is much larger than this estimate. One possible reason, which will be treated in Chap. 10, is the effect of finite temperature, which would increase the dispersion in deformation. Or there may be many nearly degenerate and shape isomers, which would have individually narrow resonances but show a broadened distribution in an ensemble.

9.4   Optical model of configurational damping

We now turn to a damping mechanism in which the collective excitation, considered as a particle–hole state, mixes with more complicated configurations such as 2-particle–2-hole states. This is an important mechanism for certain nuclear excitations, such as the giant quadrupole vibration. This mode has a very small Landau damping, but it is degenerate in energy with many other excited states of the nucleus. The residual interaction mixes the states, and we could attempt to calculate the width directly using Fermi's Golden Rule.

However, much is known empirically about the coupling of single-particle motion to the other internal degrees of freedom in a nucleus, and the damping of particle states can more reliably estimated from empirical considerations. The damping of the single-particle motion is described in a coarse way by the optical potential, which has an imaginary part W that is associated with the coupling to more complicated states. This potential is rather well known for continuum single-particle motion: scattering experiments, especially elastic scattering, directly determine the potential field distortion of the particle wave function. The real part of the potential produces a phase shift in the transmitted wave function of the particle, while the imaginary potential attenuates the particle wave function. Both kinds of distortion affect the scattered wave function of the final state.

In analogy to optics, and to discuss some of the results of the nuclear optical model in simple terms, we describe the single-particle motion in terms of the plane wave e^{ikr}, but allowing the momentum to take a complex value,

$$k = k_R + ik_I = (2m(E + V + iW))^{1/2} \qquad (9.9)$$

with $k_I > 0$. The wave function is then given by

$$e^{ikx} = e^{ik_R x} e^{-k_I x}$$

which decreases exponentially as x increases. The mean free path of the particle is the decay length for the probability density, and is given by

$$\lambda = \frac{1}{2k_I} \ .$$

Correspondingly, the mean lifetime τ for the single-particle state is given by

$$\tau = \lambda/v_g$$

where $v_g = de_k/dk$ is the group velocity associated with the particle motion. If we now assume that $W << E + V$ and expand the square root in eq. (9.9), we obtain a relation between τ and W,

$$\tau = \hbar/2W \ .$$

This relation could also be obtained more directly by considering the time-dependent Schrödinger equation for a particle in an imaginary potential. The exponential decay of the wave function implies that the strength for a given particle state is distributed in energy as

$$\frac{1}{(e - e_k)^2 + (\Gamma^{\downarrow}/2)^2}$$

in terms of the spreading width parameter $\Gamma^{\downarrow}$ given by

$$\Gamma^{\downarrow} = 2W \ .$$

This relation can be extended to a finite system where W is a function of radius r. Then we simply take the average weighted by the probability density of the particle,

$$\Gamma_p^\downarrow = \int |\phi_p|^2 W(r) d^3r \ . \tag{9.10}$$

Note that this reasoning applies equally well to the propagation of a particle above the Fermi surface or a hole in the Fermi sea.

The systematic behavior of the nuclear optical potential may be determined by fitting the optical parameters to a large body of data. An example of this kind of analysis is the global potential for neutron scattering below 30 MeV due to Rapaport et al. (1979). The associated absorptive potential is surface peaked below 15 MeV scattering energy, and increases linearly with energy. Above 15 MeV the absorption changes to a volume form. Various arguments have been advanced attempting to justify a stronger absorption in the surface region than in the interior of the nucleus. Two aspects of the surface physics must be considered. First, there are strong collective excitations of the nucleus that are associated with density changes in the surface region. These modes would only couple strongly to a particle on the surface. Second, the interaction between particles is reduced in the nuclear interior because the Pauli principle blocks final states in particle–particle scattering. The Pauli blocking is less important in the low-density surface region, and so the interaction is effectively stronger there.

We quote here the specific parameterization of the Rapaport potential, taking only the isospin-independent absorptive part:

$$iW(r) = 4iW_D \frac{df(x)}{dx}$$

$$f(x) = (1 + e^x)^{-1}, \quad x = (r - R)/a_0$$

with

$$W_D = 4.28 + 0.4E \quad \text{MeV}$$

for neutron bombarding energies $E < 15$ MeV, and $r_0 = 1.295$ fm and $a_0 = 0.59$ fm.

This is for positive neutron energies; we also need the potential for negative energies down to the Fermi energy, which is at about -8 MeV. Infinite Fermi liquid theory predicts a quadratic dependence of the damping on $\Delta e = |e - e_f|$, the energy above the Fermi level. However, there are reasons to prefer a linear dependence[†], so we shall take a simple form

$$W_D \approx 0.5\Delta e \ .$$

We can now make a rough estimate of $\Gamma^\downarrow$, approximating the integral in eq. (9.10) by half the interior W. This leads to

$$\Gamma^\downarrow = 2 < W > \approx 0.5\Delta e \ . \tag{9.11}$$

Let us now apply this to the damping of particle-hole states In a naive model, we assume that the particle and hole decay independently of each other. Then the total width is just the sum of the individual widths. Because the individual widths depend linearly on energy, the total is the same as that for a single particle at the total energy of the excitation,

$$\Gamma^\downarrow_v(\hbar\omega_v) \approx \Gamma^\downarrow_p(\hbar\omega_v/2) + \Gamma^\downarrow_h(\hbar\omega_v/2) \sim \Gamma^\downarrow_p(\hbar\omega_v) \approx 0.5\hbar\omega_v \ . \tag{9.12}$$

In Fig. 9.9 we display the estimate given by the above expression, and compare it with the experimental data on the giant dipole resonance. The peaks in the observed widths correspond to deformed nuclei ($\Gamma \approx$ 8–10 MeV) while the lower values are for spherical nuclei ($\Gamma \approx$ 5–6 MeV). Although the theoretical curve provides an average account of the observations, it overpredicts by 50% the damping width of the giant resonances of spherical nuclei. However, our expression can only be valid for spherical nuclei, where the optical potential is a purely radial function. The estimate in this section omits significant coherence effects, which we discuss in the next section.

A similar model has sometimes been applied to discuss the damping of the Mie plasmon in metals. One takes the single-particle damping from other considerations, in this case the conductivity in the Drude model. We will return to this model in Chap. 10.

9.5 Other internal degrees of freedom

In the last two sections we considered spreading due to internal couplings in two ways. We first examined a particular mode of

---

[†] The empirical spreading of bound levels is closer to linear than quadratic in ΔE, and there is also a theoretical argument by Esbensen and Bertsch (1984a).

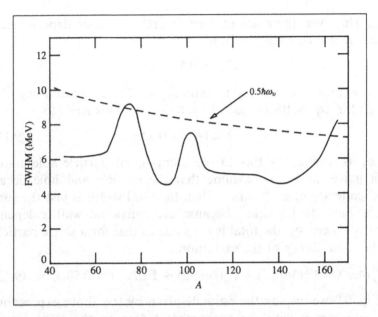

Fig. 9.9.   Width of giant dipole resonance. Dashed line is empirical, and smooth line is eq. (9.12).

coupling in detail, namely quantum quadrupole deformations. In the last section, we approached the question in a very phenomenological way, using empirical data on the spreading of single-particle states. In this section, we wish to outline a more general framework which includes these two models as special cases. When we derived the RPA theory, we considered only the parts of the interaction that left a single-particle hole excitation in the system. Parts of the interaction that simultaneously create additional particle-hole excitations or vibrational excitations were omitted. We can include all these additional pieces of the interaction into RPA using perturbation theory[†]. For example, the interaction matrix element coupling one-particle one-hole states to two-particle two-hole states, written as $<ph|v|p'p''h'h''>$, will give a spreading width if the density of final states is high enough to apply Fermi's

---

[†] More formal theory for including additional degrees of freedom in RPA is reviewed by Bertsch et al. (1983) and by Wambach (1988).

Golden Rule,

$$\Gamma^{\downarrow}_{2p2h} = 2\pi \sum_{2p2h} <ph|v|2p2h>^2 \delta(E_{2p2h} - E_{ph}) \ .$$

In the same way we could consider the coupling to a particle-hole plus vibrational state.

We will not go into great detail in the evaluation of these matrix elements but just make some general points to make contact with the previous section. In the nuclear case, the nucleon degrees of freedom are sufficient to describe the system, but it can be convenient to introduce particularly collective RPA excitations as independent vibrational degrees of freedom. The important excitations are low frequency surface-peaked modes, and we can approximate the coupling of these modes to the single-particle motion with the matrix element

$$<p|v|p',L>= \frac{\beta_L R_0}{\sqrt{2L+1}} \int d^3r \phi_p^*(\vec{r}) \phi_{p'}(\vec{r}) \frac{dV(r)}{dr} Y_{LM}(\hat{r}) \ .$$

These modes are characterized by a multipolarity L. The low frequency modes with $L = 2, 3$ are especially important, as we saw in Chap. 6.

To derive the model of Sec. 4 we would assume that the single-particle optical potential arises from this coupling. This would require

$$\Gamma^{\downarrow}_p = 2\pi \sum_{p',L} <p|v|p',L>^2 \delta(e_{p'} + \hbar\omega_L - e_p) \ .$$

On the other hand, the spreading width of a particle-hole state is

$$\Gamma^{\downarrow}_{ph} = 2\pi \sum_{p'h',L} \left(<p|v|p',L> \delta_{hh'} + <h|v|h',L> \delta_{pp'} \right)^2 \times$$

$$\delta(e_{p'} + e_{h'} + \hbar\omega_L - e_p - e_h) \ . \tag{9.13}$$

If the square in this equation is expanded, there will be two terms that are just the squares of the particle and the hole matrix elements respectively. From the previous equation, this is nothing more than the two single-particle widths. But there will also be a cross term, which could change the result from the independent decay picture. As we have written the above equation, the cross term plays only a very minor role because only the final state with $p = p'$ and $h = h'$ is affected. But if we consider a linear combination of particle-hole states, the cross term could be important.

We can see the effect of the interference in the coupling to the quadrupole surface vibrations. A single-particle state of high multipolarity will be split by the deformation. This can be estimated in the Fermi gas model, assuming that the wave function scales with the deformation. A state of momentum $\vec{k}$ then has a kinetic energy

$$e = \frac{\hbar^2}{2m}(k_z^2 e^{-4\epsilon} + k_x^2 e^{2\epsilon} + k_y^2 e^{2\epsilon}) \ .$$

We expand this to lowest order in ϵ and evaluate the dispersion at the Fermi surface, averaging over directions of $\vec{k}$. The result is

$$\sigma_p = <\Delta e^2>^{1/2} = \frac{4}{\sqrt{5}} e_f \epsilon$$

which may be expressed in terms of the deformation variable β as (cf. eq. (B.6))

$$\sigma_p = \frac{e_f \beta}{\sqrt{\pi}} \ .$$

Let us compare this with the width of the dipole state, calculated in Sec. 9.3. The dispersion there is given by eq. (9.7), which is also linearly dependent on the deformation. However, the coefficient is smaller than the sum of the particle and hole dispersions coming from the same source. In fact the ratio may be expressed

$$\frac{\sigma_D}{2\sigma_p} \approx 0.4 \frac{\hbar \omega_D}{e_f} \ ,$$

which amounts to about $1/5$. Thus we cannot simply add dispersions of particle and hole in a collective excitation.

The reason is of course that the particle and hole are coherent; the interference terms in eq. (9.13) tend to cancel. According to calculations, the cancellation should play a significant role in coupling to density vibrations as well.

10
Thermal effects

In this chapter we consider how oscillations are affected by internal excitation energy in the system. From a theoretical point of view, the main effect of the excitation energy is a broadening of the resonances, as we shall see. Empirically, one observes this behavior for the nuclear giant dipole resonance. Significant effects are also expected in the Mie resonance of metal clusters, but so far experimental techniques have not permitted measurement of the dependence on excitation energy. It is convenient and common to discuss internal excitation effects in terms of the temperature of the system, but the concept of temperature in a finite system must be used with care. So before beginning the study of oscillations, we first outline the meaning of temperature and how it is measured in a finite system. The statistical mechanical definition of temperature of a system at excitation energy E is

$$T = \left(\frac{1}{\rho_A}\frac{d\rho_A}{dE}\right)^{-1},$$

where ρ_A is the density of states of the system (per unit energy). In typical situations the density of levels is very high compared to the energy resolution with which one can specify the state of the system, and the above definition is useful and well determined. In the case of nuclei, Bethe's Fermi gas formula for the level density may be written[*]

$$\rho_A = \frac{e^{2\sqrt{aE}}}{\sqrt{48E}} \tag{10.1}$$

---

[*] There are a number of formulas in use like eq. (10.1), differing only in the prefactor. See Engelbrecht and Engelbrecht (1991) for a discussion.

172   10 *Thermal effects*

The empirical value of the parameter a at low and moderate excitation is approximately

$$a \approx \frac{A}{8} \text{ MeV}^{-1} \ .$$

For a nucleus of mass number $A = 200$ at an excitation energy of 50 MeV, the formula gives about 10^{28} states per MeV. For small metal clusters, the density of states is dominated by the molecular vibrational degrees of freedom, since the electronic degrees of freedom are hardly excited at accessible temperatures. The density of vibrational states is given by the following expression in the classical harmonic approximation,

$$\rho_A = \frac{E^{(3N-7)}}{(3N-7)!} \Pi_i^{3N-6} \frac{1}{\hbar \omega_i}.$$

Here ω_i are the vibrational frequencies of the normal modes. As an example here, we evaluate this for the Na$_{20}$ cluster (N=20), taking an average vibrational frequency of $\bar{\omega} = 10^{13}$ s^{-1} (cf. Sect. 2.1). This yields for the density of states $\rho_A \approx 10^{73}$ eV^{-1} at an excitation energy of 3 eV. In both these situations, the level density is so high that it is surely a smooth function for any energy resolution of physical interest.

In the theory of finite systems, temperature is useful for describing statistical processes in which energy is exchanged between simple degrees of freedom and the internal degrees that need only be characterized by the level density. An important process of this type is the evaporation of a particle from a finite system. Statistical reaction theory gives a formula for the decay of an excited system whose internal degrees of freedom are in equilibrium,

$$W(i) = \frac{N_i}{2\pi\hbar\rho_A} \ . \tag{10.2}$$

Here i labels a final state of the daughter system, N_i is the number of open channels or transition states leading to that daughter state, and ρ_A is the level density of the decaying system. As an example, we calculate the statistical decay rate of the compound nucleus ^{208}Pb at the energy of the giant quadrupole resonance at 11 MeV. At this energy, there are about 10^3 states per MeV with the same quantum numbers as the giant quadrupole. There are about 5 open channels for particle decay, giving a decay rate $W \approx 10^{18}$ s^{-1}.

In chemistry, this formula is known as the RRKM theory, for the contributors Rice, Ramsperger, Kassel and Marcus (Hase (1976)). The theory was actually first applied to the rate for nuclear fission, by Bohr and Wheeler (1939). According to eq. (10.2), each final level is equally likely to be populated if it is energetically allowed and the number of channels involving that state is the same. Thus the energy of the system is apportioned between kinetic energy of the evaporating particle and excitation in the daughter system solely on the basis of the density of levels of the daughter system. The more energetic the emitted particle, the lower will be the density of states in the daughter system. The evaporation rate for a particle with energy between e and $e + de$ would be given by

$$W(e)de = \frac{<N> \rho_{A-1}(E - E_B - e)}{2\pi\rho_A(E)}de \qquad (10.3)$$

where E_B is the binding energy of the particle, and $< N >$ is the average number of open channels per state of the daughter system[†] Thus $W(e)$ will be approximately exponential if ρ_A is, which will be the case if there is a large number of degrees of freedom.

In the case of nuclear decays by neutron emission, the energy spectrum of the neutrons often follows an exponential law over several decades, reflecting the exponential growth in the level density of the daughter nucleus. This is illustrated with a spectrum in Fig. 10.1. The nucleus Ta was excited to 40 MeV, and the points show the neutron emission cross section, differential with respect to energy. For lower energy neutrons, this cross section is independent of angle and exponentially decreasing. From the slope of the exponentially falling portion, we would infer a temperature of 1.4 MeV. Studies of the giant dipole resonance have been carried out up to temperatures of the order of 3–4 MeV, and will be discussed in the next section.

In the case of clusters, such detailed evaporation spectra are not yet available. However, the statistical theory can still be used to infer temperature or the excitation energy of clusters in

---

[†] It is convenient to express $<N>$ in terms of the cross section for the inverse reaction, and then one obtains the Weisskopf formula for the evaporation spectrum,

$$W(e) = \frac{2m}{\pi^2}\sigma e \rho_{A-1}(E - E_B - e)/\rho_A(E) \ , \qquad (10.4)$$

where σ is the cross section for the inverse reaction.

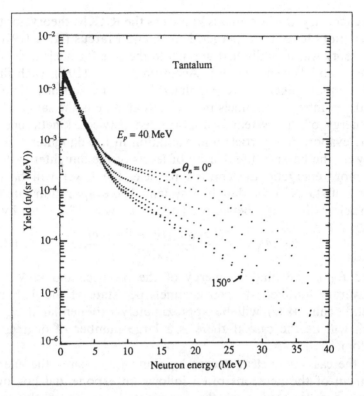

Fig. 10.1. Neutron evaporation spectra from the reaction p+Ta with 40 MeV protons, from A. Galonsky (private communication).

a beam. When the clusters are produced by expansion of a hot gas into vacuum, they will evaporate until the lifetime for the next evaporation is comparable to the transit time of the beam. A convenient formula for the total decay rate may be obtained by integrating eq. (10.3) over particle energy e, and using the approximation $\rho(E - e) \approx \rho(E)\exp(-e/T)$. This yields the simple formula for the decay rate W,

$$W = \frac{\sigma}{\pi^2}m\omega^3 e^{-E_B/T}/T \ ,$$

where σ is the cross section for a free atom to stick to the cluster. With beam transit times of the order of a millisecond, and a binding energy of 1 eV, the temperature of oven-produced sodium clusters is in the range 300–400 K.

10.1 Thermal line shifts

We now take up the question of how collective oscillations are affected by excitation energy. The frequency of the oscillation could shift, and its width could change. We consider these possibilities in turn.

As we have discussed in previous chapters, the centroid frequency of a collective oscillation is a macroscopic quantity depending on an inertia and a restoring force with respect to the motion. Both these quantities may depend on excitation energy but the dependence will be weak. For example, in the case of the Mie resonance, the dipole inertia depends only on the mass of the particles and their number, so it is completely independent of the internal state of the system. The restoring force, on the other hand, depends on the size of the system, decreasing in strength as the system expands. However, the estimated magnitude of this effect is small. Bulk sodium changes density by 3% in being warmed from 100 K to room temperature. According to the Mie formula, this would change the frequency of the resonance by 1.5%, a negligible shift.

Similar considerations may be applied to the giant dipole resonance in nuclei. The dipole resonance frequency again depends on the size of the system. To make a simple estimate of the magnitude of the frequency shift, let us assume that the excitation energy produces a pressure to expand the nucleus as in a Fermi gas, where the pressure is due to the increased speed and momentum of the particles, and is given by

$$\Delta P = \frac{2}{3}\rho\Delta E \ ,$$

where ΔE is the excitation energy per particle. The change in volume of the system may then be estimated from the compressibility modulus as

$$\frac{\Delta V}{V} = \frac{9\Delta P}{K\rho_0} = \frac{6\Delta E}{K} \ .$$

To make a numerical estimate, assume an excitation energy per nucleon of 2 MeV, and a compressibility modulus of $K = 200$ MeV. The increase in volume is only 6%, implying an increase in radius of 2%. The change in dipole frequency according to eq. (5.13) would be the same, a completely negligible effect.

10.2   Thermal line broadening

While average frequencies are not expected to shift much with
excitation energies, there are several thermal effects which can
broaden the resonances. The Landau damping will increase because of the partial occupation of single-particle orbitals. This
would allow more single-particle transitions with energies in the
resonance region. In fact, making use of the Fermi gas model one
can calculate the temperature dependence of the single-particle
velocity. Eq. (9.7) then becomes

$$\Gamma_L = \frac{\hbar v_f}{R} \left[1 + \frac{\pi^2}{6} \left(\frac{T}{e_f} \right)^2 \right]$$

to lowest order in the small parameter (T/e_f). Consequently,
Landau damping is expected to have a weak dependence on temperature.

Thermal fluctuations in the density and the shape of the cluster
can also broaden the resonances. This is apparent for the Mie
resonance and the giant dipole resonance. The resonance formulas
depend explicitly on the shape of the system, and thermal shape
fluctuations will give rise to a splitting of the resonance. When
averaged over the fluctuations, the splitting will appear to be a
broadening. This may be calculated if one knows the energy cost
for a deformation. In this section we will give a simple estimate,
assuming that the shape changes take place slowly compared to
the decay time of the resonance.

Before proceeding, we mention that the quantal fluctuations
discussed in Sect. 9.5 seem to be independent of temperature
(Bortignon et al. 1986, Nguyen 1989, De Blasio et al. 1992).

Empirical line broadening of the giant dipole

The giant dipole resonance in nuclei has been studied by measuring
the spectrum of gamma rays emitted by highly excited nuclei. If
the nuclear levels are populated statistically, the dipole emission
rate can be related to the absorption cross section by a formula
similar to eq. (10.4). The main feature of the dipole spectrum
will be an exponential fall-off due to the exponential decrease in
final state level density, but the absorption cross section appears
as a modulating factor. A typical spectrum of gamma rays is

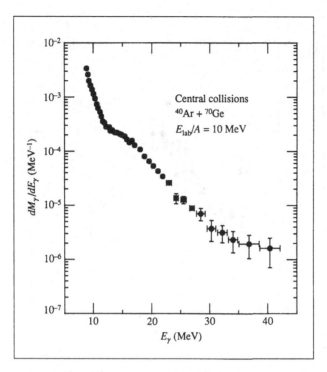

Fig. 10.2. Spectrum of gamma rays from the fusion reaction ^{40}Ar + ^{70}Ge at 10 MeV per nucleon bombarding energy, as reported by Bracco et al. (1989).

shown in Fig. 10.2. The excited nucleus was produced by a heavy ion collision, in this case ^{40}Ar+^{70}Ge, at a bombarding energy of 400 MeV. Thus the compound nucleus starts with a mass of $A = 110$ and an excitation energy of about 250 MeV. The gamma ray spectrum falls off very steeply at first, then has a shoulder and more gradual fall-off at higher energies. The spectrum is explained with a cascade process, in which all particle emissions compete, and the nucleus cools as the particles are emitted. The low energy gamma rays come primarily from the cool evaporation residue left at the end of the cascade when particle emission is no longer energetically allowed. The higher energy gamma rays come primarily from earlier stages when the nucleus is hot, and the shoulder in the curve is ascribed to the giant dipole. The very high energy gamma rays are associated with bremsstrahlung before the compound nucleus is formed.

The giant dipole is more evident if the data is presented with

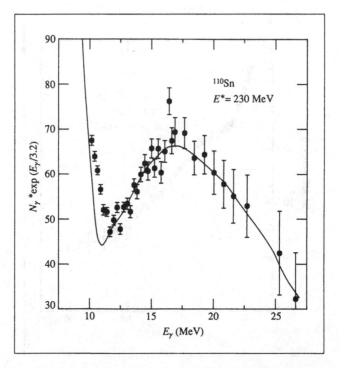

Fig. 10.3.   Gamma ray spectrum from Fig. 10.2, with bremsstrahlung subtracted out and divided by a statistical factor exp(-E_γ/3.2 MeV).

the exponential level density factor in the statistical decay formula divided out. The previous data is presented this way in Fig. 10.3. From this data, it was deduced that the giant dipole in ^{110}Sn has a mean energy of 16 MeV and a width (FWHM) of 13 MeV. The effective temperature in the exponential level density factor was taken as $T = 3.2$ MeV.

Experiments of this type show that the giant dipole does not shift significantly in frequency, but it becomes broader in a spherical nucleus that is excited. Fig. 10.4 shows the systematics of the dipole in Sn, and the increase is quite apparent.

Role of shape fluctuations

The quadrupolar fluctuations in shape are a likely source of the line broadening observed in the last section. The effect can be easily estimated, using the theory of Chap. 5 for the shape dependence of the dipole frequency. A good measure of the line broadening, as

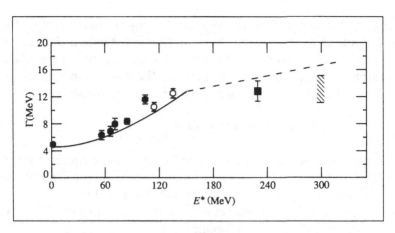

Fig. 10.4. Width (FWHM) of the giant dipole resonance in ^{110}Sn as a function of excitation energy. Data is compiled from various sources including Gaardhøje et al. (1986), Chakrabarty et al. (1987), and Bracco et al. (1989).

will be discussed in the next section, is the mean square deviation of the frequency from the center. We want to find the average of this over an ensemble of thermal states. From eq. (5.14) we find

$$< (\omega_i - \omega_D)^2 > = 2 < \epsilon^2 > \omega_D^2 \qquad (10.5)$$

We next need to find the mean square deformation in a thermal ensemble. We express the free energy of the system F as a function of the collective degrees of freedom, and evaluate thermal average of some quantity X as

$$<X> = \frac{1}{Z} \int X(\epsilon) e^{-F(\epsilon)/T} d\epsilon$$

where $Z = \int \exp(-F/T) d\epsilon$ is the normalization factor. We will take the free energy F to be the liquid drop energy of the nucleus. The surface energy of a deformed sphere may be calculated using eq. (6.6). Changing variable there from deformation length d_2 to $\epsilon = \sqrt{5/16\pi} d_2/R_0$, we have

$$F(\epsilon) = \frac{32\pi}{5} \gamma R_0^2 \epsilon^2 \quad .$$

The integral is then a simple Gaussian and the fluctuation in frequency is found to be

$$< (\omega_i - \omega_D)^2 >_\epsilon = \frac{5T}{32\pi R_0^2 \gamma} \omega_D^2 \quad .$$

There is one more fact to take into account, namely that the quadrupole deformations encompass five independent degrees of freedom, one for each of the five M-states of the quadrupole operator. The effect of this is to increase the dispersion by a factor 5. The final result for the r.m.s. dispersion in frequency is

$$< (\omega_i - \omega_D)^2 >^{1/2} = \omega_D \left(\frac{25T}{32\pi R_0^2 \gamma} \right)^{1/2} .$$

Note that this increases in the excited nuclei as the square root of the temperature. This is a direct consequence of the Boltzmann factor defining the probability of various shapes. For a numerical estimate, we consider the nucleus ^{110}Sn at a temperature of 2 MeV. In this nucleus, the dipole resonance is at 16.5 MeV, and $R_0 \approx 5.75$ MeV. The variance in the frequency is then found to be

$$< (\omega_i - \omega_D)^2 >^{1/2} \approx 2.1 \text{ MeV}.$$

As will be discussed in the next section, the line shape should be Gaussian. The width is then given by eq. (9.2), which gives $\Gamma = 2.4 < (\omega_i - \omega_D)^2 >^{1/2} \approx 4$ MeV. This is as large as the width in the cold nucleus and thus is readily observable in the experiments.

However, it is important to mention that the shape distortion in the excited nuclei such as were measured in Fig. 10.4 also arises from their rotation. The nuclei are produced by heavy ion collisions that brings in considerable angular momentum. In the liquid drop model, the deformation produced by the rotational motion is approximately given by $\beta = 4.1L^2 A^{-7/3}/(1 - 0.02Z^2/A)$ (cf. App. B). To give a numerical example, the important angular momenta associated with the ^{110}Sn measurement are in the range $L \approx 25$. The deformation is then $\beta \approx 0.08$ and the contribution to the total width may be estimated from eq. (9.8) and (9.2) as $\Gamma_{rot} \approx 1.3$ MeV. This is clearly a significant part of the total.

The shape fluctuation at finite excitation will also affect the Mie plasmon in alkali metal clusters. The dependence on the frequency is now given by eq. (5.10). The liquid drop model is not reliable for the deformation dependence of the free energy, but it may serve as a qualitative guide. Then the width formula is

$$\Gamma = 2.3 \sqrt{\frac{9T}{32\pi R_0^2 \gamma}} \omega_M .$$

As an example, we consider the sodium cluster Na$_{20}$. The bulk surface tension of sodium is given by $\gamma = 0.0125$ eV/Å^2, and the radius of the cluster is given by $R_0 = 5.7$ Å, assuming the same density as in the bulk. The resulting width at room temperature is predicted to be $\Gamma = 0.4$ eV, which happens to be close to the observed width. To make a convincing case that this mechanism is operative, one should verify the dependence on size of the cluster and on temperature. Of course, for much larger clusters other damping mechanisms discussed in Chap. 9 can become more important.

Another finite-temperature mechanism

A very simple model is sometimes employed in metal physics to describe the electron couplings in a phenomenological way, the Drude model. Here it is assumed that the electrons have a lifetime τ. This may be put into the response as an imaginary contribution to the electron energy, $Im(e) = i\hbar/2\tau$, giving the classical Mie plasmon has a Lorentzian form, eq. (2.30), with a width

$$\gamma = \frac{1}{\tau}.$$

The lifetime τ is estimated from the bulk conductivity of the metal, using the formula $\tau = m\sigma/ne^2$, where σ is the conductivity. For sodium at room temperature, $\tau \approx 3 \times 10^{-14}$ sec, which implies a width of $\Gamma \approx 20$ meV. This is much smaller than the measured widths in small clusters, $\Gamma \approx 0.3$ eV. Of course, the model is quite unrealistic anyway since it ignores the dependence of electron lifetime on the electron energy. At the energy of the Mie plasmon, the phase space for damping into other modes is much larger, with additional degrees of freedom possible involved.

10.3 A general theory

In the previous section, we discussed line broadening in the adiabatic limit, where the other degrees of freedom could be considered to be frozen during the decay of the oscillation. In principle, decay could occur also under circumstances where the thermal motion is comparable or faster than the decay lifetime. A more general theoretical framework is needed to describe line broadening in this situation, which we will present following Kubo, Toda and Hashitsume (1978). We treat the motion as a **stochastic process,** which means that we consider a time sequence of random variables perturbing the collective motion, and then average over an ensemble of systems perturbed in that way. Let us consider an oscillator whose frequency is ω_0 in the absence of thermal perturbations. The effect of the finite temperature will be to induce changes in the frequency, $\omega(t) = \omega_0 + \omega_1(t)$, where $\omega_1(t)$ is the random perturbation. The observed spectrum will be broadened, and the question is how the broadening is related to $\omega_1(t)$. This kind of problem is often encountered in the theory of spectral line shapes in atoms, molecules, and nuclear spins in matter.

The precise history of the perturbation differs from one member of the ensemble to the next, so the whole problem must be formulated in terms of probabilities. We imagine setting a system into vibration, and recording the displacement as a function of time. Repeating the measurement many times leads to readings $x_1(t), x_2(t), ..., x_N(t)$ which are all different. Each of the observed series $x_j(t)$ is a sample of a statistical ensemble. Making the number of readings N very large, one should be able to find empirically the distribution law obeyed by the stochastic variable $x(t)$.

Returning to our example of the thermally-perturbed oscillator, the basic equation of motion is

$$\frac{dx(t)}{dt} = i(\omega_0 + \omega_i(t))x(t) \ . \tag{10.6}$$

The solution to this equation is

$$x(t) = x_0 e^{i\omega_0 t} \exp\left(i \int_0^t \omega(t')dt'\right) \ .$$

The quantity x_0 is the initial value of x. If $\omega_1(t)$ is stochastic, then $x(t)$ is the stochastic process defined by this equation. In what follows we want to calculate the average intensity of the spectrum as a function of frequency. This is the **power spectrum** of the system. To define the power spectrum, we first expand the function $x(t)$ in a Fourier series

$$x(t) = \sum_n a_n e^{i\omega_n t} \ .$$

The power spectrum is then given by

$$I(\omega) = \sum_n <|a_n|^2> \delta(\omega - \omega_n) \ .$$

The power spectrum of a stochastic process may be calculated from the correlation function between the observed values of

$x(t)$ at different tome points. Since there are no transients to be considered, we may take the first time point at zero. The correlation function for eq. (10.6) is

$$<x^*(0)x(t)>= x_0^2 e^{i\omega_0 t} <\exp\left(i\int_0^t \omega_1(t')\right)dt'> \quad .$$

The brackets appearing in this equation represent ensemble averages, i.e.

$$<f(x(t))>= \frac{1}{N}\sum_i^N f(x_i(t)) \quad .$$

The Fourier transform of the correlation function defined above gives the power spectrum,

$$I(\omega) = \frac{1}{2\pi}\int_{-\infty}^{\infty} <x^*(0)x(t)> e^{-i\omega t}dt, \tag{10.7}$$

This is known as the Wiener-Kintchine theorem, and may be easily proved by using the Fourier expansion of the stochastic variable $x(t)$. Because of the stochastic nature of the variable, the different a_n are statistically independent, i.e.

$$<a_n a_n^*>= \delta(n,n') <|a_n|^2> \quad .$$

The correlation then becomes

$$<x^*(0)x(t)>= \sum_n <|a_n|^2> e^{i\omega_n t}$$

and the theorem follows immediately.

We now introduce the most important assumption of the theory, that the process is Gaussian. This means that for an arbitrary function $\xi(t)$ the ensemble average of the exponentiated integral over the variable may be calculated from the exponential of the integral over the average,

$$<\exp(i\int_0^t \xi(t')\omega(t')dt')>$$

$$= \exp\left(-\frac{1}{2}\int_{t_0}^t dt_1 \int_{t_0}^t dt_2 <\omega(t_1)\omega^*(t_2)> \xi(t_1)\xi(t_2)\right) \quad . \tag{10.8}$$

One can then write

$$<x^*(0)x(t)>= \exp\left(-\frac{1}{2}\int_0^t dt_1 \int_0^t dt_2 <\omega_1(t_1)\omega_2(t_2)>\right) \quad .$$

Because the ω_1 are stochastic, the average on the right hand side depends only on the time difference and can be written as

$$< \omega_1(t_1)\omega_2(t_2) >= (\delta\omega)^2\Psi(t_1 - t_2)$$

where $\Psi(t)$ is some function that decays to zero for large times and is unity at $t = 0$. The $(\delta\omega)^2$ is the mean square spread of frequencies induced by the random modulation. Substituting the above relation into eq. (10.7) and making the change of variable $t' = t_1 - t_2$ leads to

$$< x^*(0)x(t) >= \exp\left(- (\delta\omega)^2 \int_0^t (t - t')\Psi(t')dt' \right) .$$

Finally, it is common to approximate $\Psi(t)$ by an exponential decay,

$$\Psi(t) = e^{-t/\tau_c}$$

where the parameter τ_c is the decay constant. Then the correlation function becomes

$$< x^*(0)x(t) >= \exp(-\alpha(\frac{t}{\tau_c} - 1 + \exp(-t/\tau_c))) . \qquad (10.9)$$

with

$$\alpha = \delta\omega\tau_c .$$

The parameter α measures the degree of adiabaticity of the motion. If α is large, the thermal motion is slow compared to the decay rate of the vibration, and the adiabatic theory of the previous section applies. When α is small, the thermal motion is rapid, and the power spectrum of the vibration is quite different. This may be seen in Fig. 10.5, showing the Fourier transform of the correlation function, eq. (10.9). The linewidth narrows as the parameter α decreases, an effect well-known in nuclear magnetic resonance and called **motional narrowing**. We see that temporary shift of the frequency is significant in the spectrum only if its duration is comparable to or larger than $\tau = 2\pi/\delta\omega$. If the duration is shorter than τ, the modulation is averaged out and is difficult to observe.

The line shape of the power spectrum has a simple form in the two limits, $\alpha << 1$ and $\alpha >> 1$. When α is large, one can make a power series expansion of $\Psi(t)$ and the correlation function reduces to the form

$$< x^*(0)x(t) >= \exp(-\frac{1}{2}\delta\omega^2 t^2) .$$

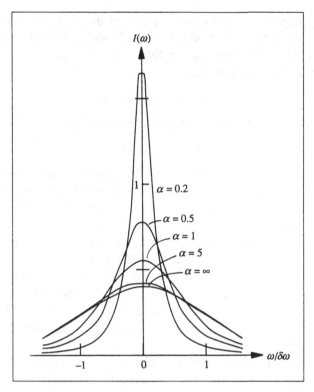

Fig. 10.5.   Power spectrum associated with the correlation function eq. (10.9).

This is a Gaussian, which also has a Gaussian Fourier transform. Thus, the power spectrum is predicted to have a Gaussian shape for large values of α. In the opposite limit, $\alpha \ll 1$, the correlation persists for long times. In making the Fourier transform of eq. (10.9), contributions from $t \gg \tau_c$ will dominate, allowing one to neglect the e^{-t/τ_c} term in eq. (10.9). Thus the correlation function behaves as

$$< x^*(0)x(t) > \sim e^{-\alpha\delta\omega t}$$

The Fourier transform of this correlation function is just the Breit–Wigner distribution,

$$I(\omega) \sim \frac{\alpha\delta\omega}{(\omega - \omega_0)^2 + (\alpha\delta\omega)^2}$$

As may be seen from Fig. 10.5, the power spectrum shows a more broad-based Breit–Wigner shape for $\alpha \ll 1$.

The distribution in this limit is narrower according to one of the principal measures of the width, the full width at half maximum (FWHM). This quantity is given by eq. (9.2) in the adiabatic limit because the line shape is Gaussian,

$$\Gamma(\alpha \gg 1) \approx 2.4\delta\omega$$

In the motionally narrowed regime it is given by

$$\Gamma(\alpha \ll 1) \approx \alpha\delta\omega$$

and is obviously smaller than $\delta\omega$. Although the peak is narrower, the line shape has wings that fall off rather slowly. The second moment of the frequency distribution is infinite in the motionally narrowed regime, unlike the adiabatic regime. Thus the rapid thermal motion appears only in the wings of the power spectrum.

In the giant dipole of nuclei and the Mie resonance of metal clusters, the physical time scales would seem to favor the adiabatic limit. In the cluster case, the large mass of the ions makes the time scale for thermal motion rather long, $\hbar/\tau_c \approx 1\text{–}10$ meV. Systems with much larger observed widths, for example, the sodium cluster shown in Fig. 1.4, would not be motionally narrowed. However, cases exist where the width is small, e.g. the potassium cluster shown in Fig. 9.4, and then the adiabatic theory could not be used.

In the nuclear case, the time scale for thermal shape changes is not known very well. There is empirical information on the dynamics of the shape degree of freedom coming from the study of nucleus–nucleus collisions (Schroeder and Huizenga (1984)), and also on the competition between fission and other modes of decay in hot nuclei(Gavron et al. (1981,1987), Hinde et al. (1988) and Thoennessen et al. (1987)). From Fig. 10.4, we might suspect that motional narrowing could be responsible for the saturation in the dipole width, but this is still an open question.

Appendix A
Mean field theory

In this appendix we want to summarize the mean-field theory of the ground state, the starting point for calculating excitation properties in RPA. Quantum mechanical mean field theory is inspired by the Hartree–Fock approximation of atomic physics. The fundamental approximation in the Hartree–Fock theory of fermion systems is that the many-particle wave function is a determinant composed of single-particle wavefunctions. The Hartree–Fock theory finds the optimum determinant, in the sense that it minimizes the expectation of the Hamiltonian. Let us make this specific by taking a Hamiltonian having the usual single-particle kinetic energy operator together with a two-particle interaction,

$$H = -\sum_i \frac{\hbar^2}{2m}\nabla^2 + \sum_{i<j} v(r_{ij}) \ .$$

Then the optimum many-particle wave function is obtained by using single-particle wave functions ϕ_i that satisfy

$$-\frac{\hbar^2}{2m}\nabla^2\phi_i(r) + \int d^3r' \sum_j |\phi_j(r')|^2 v(r-r')\phi_i(r)$$

$$-\int d^3r' \sum_j \phi_j^*(r')\phi_i(r')\phi_j(r)v(r-r') = e_i\phi_i(r) \ . \qquad (A.1)$$

The e_i arises as a Lagrange multiplier in minimizing $< H >$, and is interpreted as the single-particle energy. The first interaction term may be rewritten as an ordinary potential term $V\phi_i$ with the

187

potential given by

$$V(r) = \int d^3r' n(r')v(r - r')$$

where

$$n(r) = \sum_{j=1}^{N} |\phi_j(r')|^2$$

is the density of electrons.

The second interaction term in eq. (A.1), the exchange interaction, cannot be expressed so simply. As a practical matter, the Schroedinger equation is only easy to solve for such simple potential fields. This gives a strong motivation to approximate the exchange interaction by a form that would make it fit into a potential field. In atomic physics, the following approximation was introduced by Slater:

$$-\int d^3r' \frac{e^2}{|r - r'|} \sum_j \phi_j^*(r')\phi_i(r')\phi_j(r) \approx c(r)\phi_i(r)$$

where the coefficient $c(r)$ is calculated assuming the occupied wave functions can be represented by an infinite Fermi gas in the vicinity of the point r. This yields[†]

$$c(r) = -e^2 \left(\frac{3n(r)}{\pi}\right)^{1/3} \approx -0.984 e^2 n^{1/3}$$

where $n(r)$ is the density of electrons at r. Once the Slater approximation is made to simplify the Hartree–Fock equations, it is tempting to make further changes in the interaction in order that the final energies correspond better to the actual solution of the many-particle Schroedinger equation. This approach is known as the Local Density Approximation and the Schroedinger equation with the modified potential field is called the Kohn–Sham equation[‡]. Besides the direct and exchange Coulomb interactions between the electrons, the potential includes a term $V_{ion} = -e^2 \sum_\alpha 1/|\vec{r} - \vec{R}_\alpha|$ describing the attractive interaction between the electrons and the ions, and a density-dependent contact interaction chosen to reproduce the correlation energy of an interacting electron gas. Thus

---

[†] See Heine (1985) for a derivation.

[‡] A justification for this procedure that is often cited is the density functional theorem of Hohenberg and Kohn (1964).

the Kohn–Sham equation has the form

$$-\frac{\hbar^2 \nabla^2}{2m}\phi_i + V\phi_i + V_{ion}\phi_i - 0.984e^2 n^{1/3}\phi_i + \frac{\partial E_{corr}}{\partial n}\phi_i = e_i\phi_i$$

where $E_{corr}(n)$ is the correlation energy in the interacting electron gas[§] The LDA has been very successful in solid state physics, as discussed in Lundqvist and March (1983). For example, the total energy derived from the LDA gives a good indication of the most stable state, when a choice of states is possible. This may be used to predict the crystal structure of the material, whether the metal is paramagnetic or ferromagnetic, etc. However, it should be mentioned that the LDA has definite limitations in describing electronic excitations (see Hanke et al. (1985)); LDA typically underestimates band gaps in semiconductors and insulators by 30–50%. This shows the importance of the nonlocality of the exchange interaction in describing single-particle energies.

In nuclear physics, the corresponding local density approximation has also been applied with much success (see Negele (1982)). However, the nuclear interaction is rather complicated in detail and for many purposes a more crude approach is adequate. In this spirit many calculations use the Skyrme Hamiltonian[†]. The single-particle Hamiltonian has the form

$$H = -\frac{\hbar^2 \nabla^2}{2m^*} + V,$$

the sum of a kinetic energy operator and a potential functional. The essential feature of V is that it is local, a function of r that depends only on quantities calculable from the density matrix in the vicinity of r. In the usual parameterization V is expressed

$$V(r) =$$

$$\frac{3}{4}t_0\rho(r) + \frac{3}{16}t_3\rho^2(r) + \frac{1}{16}(3t_1 + 5t_2)\tau(r) + \frac{1}{32}(5t_2 - 9t_1)\nabla^2\rho(r)$$

where we have simplified the full expression by assuming equal numbers of neutrons and protons and no partially filled shells. The first term with t_0 is associated with the basic short-ranged attraction between nucleons. By itself this would lead to a collapse

---

[§] This is often parameterized as $E_{corr} = -0.91 \log(1+11.4/r_s)$ eV, with $r_s = (4\pi n(r)/3)^{1/3}$ in atomic units.

[†] See Vautherin and Brink (1972) for details.

of the nucleus in Hartree–Fock theory. The next two terms are repulsive and serve to offset the attraction, producing nuclear saturation at some finite density. The second term with t_3 depends quadratically on ρ, and the third term depends on the kinetic energy density τ which varies as the $5/3$ power of density in the Fermi gas limit. The kinetic term in the single–particle Hamiltonian has an effective mass that depends explicitly on density,

$$\frac{1}{m^*} = \frac{1}{m}(1 + (3t_1 + 5t_2)\rho) \ ,$$

which is not appreciably more difficult to treat than the usual Schroedinger equation with a constant mass. The effective masses arise from momentum-dependent contact interactions; the momentum dependence is present in the true nucleon–nucleon interaction and is a consequence of the meson exchange character of the interaction. However, quantitatively the coefficients t_i cannot be calculated very well from the basic interaction[†].

Like LDA in solid state physics, the Skyrme model has been successful in describing the structure of nuclei and their binding energies. Since the entire interaction is phenomenological, it is not surprising that the average systematics of energies is reproduced. But it also gives the basic shell physics: magic numbers, deformations, the quantum numbers of single-particle orbitals near the Fermi surface. Also the charge distributions within nuclei are reasonably well reproduced by the mean field theory (See Fig. A.1).

For rough estimates it is useful to replace the self-consistent potential by a three-dimensional harmonic oscillator potential. The wave functions and energies are then governed by a single parameter, which we may take to be the oscillator frequency. In approximating nuclear wave functions, the oscillator is commonly determined by requiring the mean square radius of the system to match that of a uniform density sphere. The latter quantity is

$$<r^2> \approx \frac{3}{5}(r_0 A^{1/3})^2 \ .$$

The mean square radius of an oscillator wave function having n

---

[†] In fact, according to many-particle perturbation theory, the power series expansion for the density dependence should have $\rho^{1/3}$ as the variable instead of ρ.

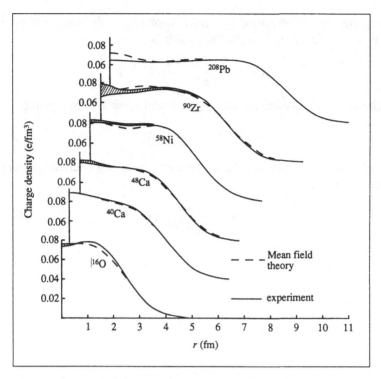

Fig. A.1. Comparison of mean field predictions for charge distributions with empirical data. The experimental uncertainty in the charge
distribution is indicated by the shaded area between the solid lines.

quanta of excitation is

$$<r^2> = \frac{\hbar}{m\omega_0}(n + \frac{3}{2}) \ .$$

We need to average this over the N particles in the system. Let
us assume that the levels are filled to quantum number n_{max} and
the spin degeneracy of the levels is g. Then the total number of
particles is

$$N = \frac{g}{2}\sum_{n=0}^{n_{max}}(n + 1)(n + 2) = \frac{g}{6}(n_{max} + 1)(n_{max} + 2)(n_{max} + 3) \ .$$

The mean square radius has a similar sum,

$$N <r^2> = \frac{g\hbar}{2m\omega_0}\sum_{n=0}^{n_{max}}(n + 1)(n + 2)(n + 3) \ .$$

Using the above relations, we may approximately solve for ω_0 in terms of N. In energy units, the result is

$$\hbar\omega_0 = \frac{5}{4}\left(\frac{6}{g}\right)^{1/3}\frac{1}{r_0^2}N^{1/3} \ .$$

In the nuclear physics context, $g = 4$ and $r_0 \approx 1.2$ fm, giving

$$\hbar\omega_0 \approx \frac{41}{A^{1/3}} \text{ MeV}. \tag{A.2}$$

For electrons in metal clusters, the corresponding oscillator energy is given by

$$\hbar\omega_0 \approx \frac{47}{r_s^2 N^{1/3}} \text{ eV}. \tag{A.3}$$

Specification of deformations

There are many ways in common use to describe the deformation of the nucleus. Bohr and Mottelson (1975) introduced an expansion of the position of the surface in terms of spherical harmonics. The distance to the surface along the direction (θ, ϕ) is expressed in terms of the deformation coordinates α_{LM} as

$$R(\theta, \phi) = R_0 \left(1 + \sum_{LM} \alpha_{LM} Y_{LM}(\theta, \phi) \right) \quad . \tag{B.1}$$

In dealing with spheroidal shapes we have $R(\theta, \phi) = R(\theta)$ and only one term in the expansion, $Y_{20}(\theta)$, needs to be taken. Defining $\beta_2 = \alpha_{20}$, the above equation reduces to

$$R(\theta) = R_0 (1 + \beta_2 Y_{20}(\theta, \phi)) \quad . \tag{B.2}$$

The major and minor axes of this spheroid are given by

$$R_z = \left(1 + \beta_2 \sqrt{\frac{5}{4\pi}} \right) R_0 \tag{B.3}$$

$$R_x = R_y = \left(1 - \beta_2 \frac{1}{2} \sqrt{\frac{5}{4\pi}} \right) R_0 \quad . \tag{B.4}$$

When $\beta > 0$, the nucleus is a prolate or cigar-shaped spheroid. Another quantity which is frequently used for spheroidal shapes is the distortion of the system, given by

$$\delta = \frac{R_z - R_x}{R_0} = 3 \sqrt{\frac{5}{16\pi}} \beta \quad . \tag{B.5}$$

Numerically $\delta = 0.95\beta$ and the two parameters are very much alike.

The scaling transformation used in the text, given by eq. (2.25), depends on a parameter ϵ. This transformation is equivalent to the β_2 transformation to first order, but unlike that parameterization it preserves density to all orders. The relation, derived by comparing the semimajor axes to first order, is

$$\epsilon = \sqrt{\frac{5}{16\pi}}\beta_2 \tag{B.6}$$

For general multipoles, we have considered displacement fields of the form

$$\vec{u} = a\nabla r^L Y_{LM}(\hat{r}) \ .$$

We compare $R_0 + u_z(R_0, \hat{z})$ with eq. (B.1) on the z-axis to express α_{LM} in terms of a,

$$\alpha_{LM} = aLR_0^{L-1} \ .$$

We have also defined a deformation length d_L which is simply related to α_{LM} as

$$d_L = R_0\alpha_{L0} \ .$$

Influence of angular momentum

In nuclei produced by heavy ion reactions, the deformation can be significantly affected by the rotational motion. We sketch here an estimate based on the liquid drop model. More detailed discussion and other references can be found in Cohen, Plasil, and Swiatecki (1974). For not too large rotational angular momentum, the preferred shape is oblate with the symmetry axis along the rotational axis. The total liquid drop energy as a function of angular momentum $\hbar I$ and deformation β is then given by

$$E = \frac{\hbar^2 I^2}{2I_r(1 + \sqrt{5/4\pi}\beta)} + \frac{1}{2}C_2^{\beta}\beta^2.$$

Here $I_r = 2AMR_0^2/5$ is the rigid moment of inertia of a sphere, and C_2^{β} is the liquid drop restoring force coefficient. The contribution of the surface tension to the coefficient is given by eq. (6.7). However, for heavy nuclei the Coulomb force is also significant and the two together give a coefficient

$$C_2^\beta = 4\gamma R_0^2 (1 - x)$$

where $x \approx 0.02 Z^2/A$ is the usual fissility parameter of the liquid drop model. To obtain the static deformation at finite angular momentum, one minimizes the above energy function with respect to β. The result is

$$\beta_{min} \approx 4 \frac{I^2}{(1 - 0.02 Z^2/A) A^{7/3}} \ .$$

We use this result in Sect. 10.2.

Appendix C
Finite nucleus compressibility

In this appendix we examine the compressibility of a finite nucleus, to see the importance of surface effects. We start from a mean field model in which the potential energy density is expressed as a power series in the particle density as in the Skyrme interactions discussed in App. A,

$$\mathscr{V} = v_1 \rho^2 + v_2 \rho^3 \ .$$

For convenience we express ρ in units of the nuclear matter saturation density which we will take as $\rho_0 = 0.162$ fm^{-3}. The coefficients v_1 and v_2 then have dimensions of energy. The connection with the Skyrme parameterization is $v_1 = 3t_0\rho_0/8$ and $v_2 = t_3\rho_0^2/16$. The other contribution to the energy is the single-particle kinetic energy, which in a Fermi gas is related to the density by

$$T = \frac{3\hbar^2}{10m}\left(3\pi^2\rho\rho_0/2\right)^{2/3} = 22.3\rho^{2/3} \ \text{MeV}. \tag{C.1}$$

Here we have used the mean square momentum of the particles $<(\hbar k)^2> = 3\hbar^2 k_f^2/5$ where $\hbar k_f = (3\pi^2\rho_0/2)^{1/3}$ is the Fermi momentum.

The parameters v_1 and v_2 can be fixed by the saturation requirements of nuclear matter. The condition that nuclear matter be in equilibrium at the density $\rho = 1$ implies that the energy is stationary there,

$$\frac{d}{d\rho}\left(T + \frac{\mathscr{V}}{\rho}\right)_{\rho=1} = 0 \ .$$

This gives the following relation between v_1 and v_2,

$$v_1 + 2v_2 + 14.9 = 0 \ \text{MeV}. \tag{C.2}$$

The second empirical condition that the model must satisfy is that the binding energy at saturation be the nuclear matter value, which is about 16 MeV per nucleon. This gives the equation

$$v_1 + v_2 + 22.3 = -16 \quad \text{MeV}$$

These equations are both satisfied with the values

$$v_1 = -61.3 \quad \text{MeV} \quad \text{and} \quad b = 23.3 \quad \text{MeV}.$$

The infinite matter compressibility may be found from eq. (7.1),

$$k = \rho\rho_0 \frac{\partial}{\partial\rho} \rho^2 \frac{\partial}{\partial\rho} E(\rho)$$

$$= 2v_1\rho_0\rho^2 + 6v_2\rho_0\rho^3 + \frac{10}{9}(22.3 \text{ MeV})\rho_0\rho^{5/3} \quad .$$

Inserting the values of v_1 and v_2 determined by the saturation conditions, the nuclear compression modulus is predicted to be

$$K = \frac{9k}{\rho\rho_0} = 18v_1 + 54v_2 + 10 \times 22.3 \approx 380 \quad \text{MeV}. \qquad (C.3)$$

We now trace how this model produces a much lower effective nuclear compression modulus in finite nuclei. We assume that the oscillation is uniform breathing with the field of eq. (7.3). We need to construct the energy as a function of the scaling parameter a to apply the Rayleigh principle. The numerator function consists of the v_1 and v_2 interaction terms together with single-particle kinetic energy from eq. (C.1). The total depends on a as

$$\mathcal{V}(a) = A\left(\frac{v_1 <\rho>}{(1+a)^3} + \frac{v_2 <\rho^2>}{(1+a)^6} + \frac{1}{(1+a)^2} < -\frac{\hbar\nabla^2}{2m} > \right) \quad .$$

Here A is the total number of nucleons in the nucleus, and the bracketed quantities are expectation values in the ground state, i.e.

$$<\rho^n> = \int d^3r\rho^{n+1}(r) / \int d^3r\rho(r) \quad .$$

The numerator function in Rayleigh's principle may then be expressed as

$$\frac{\partial^2 \mathcal{V}}{\partial a^2}\bigg|_{a=0} = A\left(6a <\rho> +30b <\rho^2> +2 < -\frac{\hbar\nabla^2}{2m} > \right) \qquad (C.4)$$

In a very large nucleus, this would be equivalent to eq. (C.3) because the expectation values would be close to the interior values*, $<\rho^n> \approx 1$ and $<-\hbar\nabla^2/2m> \approx 3\hbar^2 k_f^2/10M$. However, the fact that the v_1 and v_2 terms have opposite signs leads to a greater sensitivity to the precise values of $<\rho>$ and $<\rho^2>$ than might have been expected. For actual numbers, we take ρ to have the form

$$\rho_0 = \frac{1}{1 + \exp((r - r_0 A^{1/3})/a_0)}$$

with $r_0 = 1.2$ fm and $a_0 = 0.55$ fm. For the nucleus ^{208}Pb this yields $<\rho> = 0.73$ and $<\rho^2> = 0.63$, which is a substantial difference from unity. The last term, the single-particle kinetic energy, does not change so much and is anyway much less important. Inserting the ^{208}Pb expectation values in eq. (C.4), the nuclear compression modulus decreases from 380 to 210 MeV. Thus the finite surface effects produce an enormous change in the effective kinetic energy T. Similar changes are found with more sophisticated mean field models. Thus one cannot make a completely empirical extraction of the nuclear compression modulus from the breathing mode frequency, but must rely on theory to account for the surface effects (cf. also Brown (1988)).

---

* The coefficients are not the same in the two equations, but they can be shown to be equivalent using eq. (C.2).

Appendix D

Nuclear surface reactions

In this appendix we present a simplified theory of direct reactions that excite surface modes. A characteristic of direct reactions is that the projectile is only slightly deflected by the interaction. This allows one to describe the reaction in terms of a very limited number of degrees of freedom: the projectile-target coordinate and the excitation amplitude. In general, a nuclear projectile interacts strongly with the target and perturbation theory is not valid for describing the wave function. However, the amplitudes of the surface modes are small enough so that perturbation theory may be applied to the excitation process itself. The usual theory for this regime is the distorted wave Born approximation, which treats the projectile motion quantum mechanically. From a computational point of view the theory is rather complicated, but fortunately a simple analytic approximation is applicable under certain conditions.

For massive projectiles and high bombarding energy one may use the eikonal approximation, which assumes that the projectile moves along a straight line, say in the z-direction, and has a negligible change in momentum during the scattering. The part of the wave function at impact parameter b acquires an additional phase $\exp(i\chi)$ according to the formula

$$\chi(b) = \frac{1}{\hbar v} \int_{-\infty}^{\infty} V(\sqrt{b^2 + z^2})dz \qquad (D.1)$$

Here v is the velocity of the projectile, and V is the projectile-target interaction. We will use V in two ways. The first is as an ordinary potential, which would by itself produce only elastic scattering.

The V can also be viewed as a matrix in the space of target states. The diagonal element in the ground state is just the ordinary potential, $V = <0|\hat{V}|0>$. The interaction $\hat{V}$ also has off-diagonal matrix elements which are simply the transition potential to excite various states, $\delta V = <f|\hat{V}|0>$. Then the amplitude c to excite the state, treated perturbatively in the transition potential, is

$$c(b) = <i| \exp(i\chi(\hat{V}))|f> \approx e^{i\chi} <i|\chi(\hat{V})|f>$$

$$= \frac{e^{i\chi}}{\hbar v} \int_{-\infty}^{\infty} \delta V dz \ . \tag{D.2}$$

We next assume that the collective model can be used to describe the transition potential. Thus we take δV to have the form

$$\delta V = \frac{\beta_L R_0}{\sqrt{2L+1}} Y_{LM}(\hat{r}) \frac{dV}{dr} \tag{D.3}$$

The nuclear interaction varies very rapidly with r in the vicinity of the contact distance, and the important region of integration in eq. (D.2) is near $z \approx 0$ with b is close to the contact point R_S. Under these conditions we can express the transition eikonal factor as

$$\delta\chi = \frac{1}{\hbar v} \int_{-\infty}^{\infty} \delta V(r) dz = \frac{\beta_L R_0}{\hbar v \sqrt{2L+1}} \int \frac{dV}{dr} Y_{LM}(\hat{r}) dz$$

$$\approx \frac{\beta_L R_0}{\sqrt{2L+1}} \frac{d\chi(b)}{db} Y_{LM}(\pi/2, \phi) \ .$$

The angular distribution can be obtained from the Fourier transform of the amplitude $c(b)$:

$$\frac{d\sigma}{d\Omega} = \frac{k^2}{4\pi^2} |\tilde{c}(k \sin \theta)|^2$$

where k is the reduced wavenumber of the projectile and

$$\tilde{c}(q) = \int d^2 b e^{i\vec{q}\cdot\vec{b}} c(\vec{b}) \ .$$

The angular integration in this equation gives a Bessel function leaving the following expression,

$$\tilde{c}(q) = \frac{\beta_L R_0}{\hbar v \sqrt{2L+1}} Y_{LM}(\pi/2,0) \int_0^{\infty} b \frac{d\chi}{db} e^{i\chi} J_M(qb) \, db \ . \tag{D.4}$$

We only want to use this expression to describe roughly the angular distribution. Note first that the integrand factor $d\chi/db \exp(i\chi)$ will be large only in the vicinity of the contact point: χ is large and imaginary for small r, because of the absorption of the projection, and this causes the exponential factor to cut off the integrand at small r. But χ itself is small at large r, and the factor $d\chi/db$ cuts off the integrand in that region. Thus the amplitude peaks at values of $q \approx x_i/R_S$, where x_i are the maxima for the J_M Bessel function. We need only know the M values of the transition and the strong interaction radius R_S to get the peaks of the cross section. For quadrupole excitations, the spherical harmonics in eq. (D.4) have the values $Y_{LM}(\pi/2,0) = \sqrt{5/16\pi}, 0$, and $\sqrt{15/32\pi}$ for $|M| = 0, 1$ and 2. Thus the $M = 2$ amplitude is largest, and the peak in the differential cross section is at the peak of the J_2 Bessel function, $qR \approx 3.1$. In contrast, the monopole excitation has $M = 0$ only and peaks at $q = 0$. It is also easy to see that all excitations with odd L vanish at $q = 0$.

We finally derive a crude estimate of the angle-integrated cross section to excite a particular state, using the relation

$$\sigma \doteq \int d^2b |c(b)|^2 \ .$$

We need to assume that χ is purely imaginary, which is a reasonable approximation at high projectile energies. Then the integral for the cross section becomes

$$\sigma = 2\pi \left(\frac{d\chi_I}{db}\right)^2 \frac{(\beta_L R_0)^2}{2L+1} \sum_M |Y_{LM}(\pi/2,0)|^2 \int_0^\infty b\,db\,e^{-2\chi_I}$$

To simplify this further, first note that the average value of the modulus squared of a spherical harmonic is $1/4\pi$. Thus

$$\frac{1}{2L+1} \sum_M |Y_{LM}(\theta,\phi)|^2 = \frac{1}{4\pi} \ .$$

Next, we use one of the derivative factors to change the integration variable, $(d\chi_I/db)\,db = d\chi_I$. Since χ_I varies rapidly with b, we may take the factor b outside the integrand, replacing it by R_S, the strong absorption radius.

Finally, we assume that χ_I varies exponentially with b,

$$\chi_I(b) \approx Ae^{-b/a_0}$$

where a_0 is the diffusivity of the potential V.

This approximation is motivated by the exponential dependence of the potential on distance in the surface region, but in practice the approximation is rather rough. However, we can then replace the remaining factor of $d\chi_I/db$ in the integrand by χ_I/a_0. The integral is then reduced to an analytic form,

$$\sigma = \frac{\beta^2 R_0^2 R_S}{2a_0} \int_0^\infty \chi_I e^{-2\chi_I} d\chi_I = \frac{\beta^2 R_0^2 R_S}{8a_0} \ . \tag{D.5}$$

This expression yields estimates within a factor of two or so of more careful treatments of the eikonal approximation.

Appendix E

Numerical RPA

For readers interested in numerical RPA calculations in spherical systems, three computer programs are available from the authors. Two of the programs, RPA and JELLYRPA, use the response function in coordinate space to model the excitations of nuclei and of metal clusters, respectively. The third program, GIANT, using the configurations space matrix technique for modelling nuclear excitations. The programs can be obtained by electronic mail from bertsch@phys.washington.edu or from broglia@vaxmi.infn.it. A brief description of the essential numerical technique follows.

In the coordinate space response technique, the independent particle polarization propagator is decomposed in an expansion over spherical harmonics, which are not coupled in a spherical system. Thus

$$\Pi^0(\vec{r}, \vec{r}') = \sum_{L,M} \Pi^0_L(r, r') Y^*_{LM}(\hat{r}) Y_{LM}(\hat{r}') \ .$$

The polarization propagator for a given spherical harmonic L and frequency ω is expressed in terms of the wave functions of the occupied states ϕ_h and the single particle Green's functions $g_{j'}$ as

$$\Pi^0_L(r, r', \omega) =$$

$$\sum_{h,j'} c^2(j_h j' L)\phi^*_h(r)\phi_h(r') \Big(g_{j'}(r, r', \omega - e_j) + g_{j'}(r, r', -\omega - e_j)\Big)$$

where j_h, j' label angular momentum of the particles and $c^2(jj'L)$ is an angular factor[†]. The single-particle Green's function is ex-

---

[†] The factor is expressed in terms of the Clebsch–Gordon coefficient $(j'm'jm|LM)^2$ as:

pressed in terms of the regular and outgoing-wave solutions of the radial Schroedinger equation, u and w, as

$$g_j(r, r', e) = \frac{2mu(r_<)w(r_>)}{w(du/dr) - u(dw/dr)} \ .$$

The independent-particle polarization propagator is constructed as a matrix on a mesh in coordinate space, and then the operator manipulations demanded by eq. (4.2) are carried out as matrix operations to obtain the RPA polarization propagator. The strength function is obtained from eq. (4.4), which requires that the matrix be multiplied by vectors on both the right and left hand sides. This gives the strength function at one energy; to get the energy distribution the calculation is repeated over a mesh of energy points. Further description of these programs may be found in Bertsch (1990) and (1991).

The third program, GIANT, was written by Van Giai Nguyen to calculate the fully self-consistent nuclear RPA response with the Skyrme Hamiltonian, with extensive revisions by G. Colò. GIANT uses the particle-hole configuration representation and a harmonic oscillator basis for the single-particle wave functions. The program first calculates Hartree–Fock wavefunctions and energies associated with a Skyrme Hamiltonian, and then sets up the the RPA A and B matrices defined in eq. (4.8). The combined matrix is not symmetric and requires two steps to diagonalize with standard routines. This is done following the the procedure of Ullah and Rowe (1970). One first diagonalizes the matrix $(A - B)$, which allows one to take its square root. Next one diagonalizes the symmetric matrix $(A - B)^{1/2}(A + B)(A - B)^{1/2}$, which gives the desired eigenvalues in terms of the eigenvectors $(A - B)^{-1/2}(X + Y)$.

$c^2 = (2j + 1)(2j' + 1)(j'0j0|L0)^2/4\pi(2L + 1)$ if j, j' are orbital angular momenta and spin is ignored, and $c^2 = (2j + 1)(2j' + 1)(j'1/2j - 1/2|L0)^2/16\pi(2L + 1)$ if j, j' are spin-orbital coupled angular momenta.

References

Ahrens, J., Borchert, H., Czock, K., et al. (1972). *Nucl. Phys.* **A251** 479.

Barma, M. and Subrahmanyam, V. (1989). *J. Phys.: Cond. Matter* **1** 7681.

Barrette, J., Alamanos, N., Auger, F., et al. (1988). *Phys. Lett.* **B209** 182.

Berman, B. and Fultz, S. (1975). *Rev. Mod. Phys.* **47** 713.

Bernath, M., Yannouleas, C. and Broglia, R. (1991). *Phys. Lett.* **A156** 307.

Bertrand, F., (1976). *Ann. Rev. Nucl. Sci.* **26** 457.

Bertrand, F., Beene, J. and Horen, D. (1988). *Nucl. Phys.* **A482** 287.

Bertsch, G. (1974). *Phys. Rev.* **A9** 819.

Bertsch, G., Barranco, F. and Broglia, R.A. (1986). In *Unified Concepts of Many-Body Physics,* ed. Kuo and Speth, Vol. I (North-Holland, New York) 33.

Bertsch, G., and Hamamoto, I. (1982). *Phys. Rev.* **C26** 1323.

Bertsch, G., Bortignon, P. and Broglia, R. (1983). *Rev. Mod. Phys.* **55** 287.

Bertsch, G. and Esbensen, H. (1985). *Phys. Lett.* **161B** 249.

Bertsch, G. and Esbensen, H. (1987). *Rep. Prog. Phys.* **50** 607.

Bertsch, G.F. (1988). In *Frontier and borderlines in many-particle physics,* ed. Broglia and Schrieffer (North-Holland, New York) 41.

Bertsch, G.F. (1990). *Computer Phys. Comm.* **60** 247.

Bertsch, G.F. (1991). In *Computational nuclear physics,* ed. Langanke, Maruhn and Koonin (Springer, New York), 75.

Bes, D. and Sorensen, R. (1969). *Advances in Nuclear Physics* **2** 129.

Blanpied, G., Coker, W., Liljestrand, R., et al. (1978). *Phys. Rev.* **C18** 1436.

Blaizot, J. (1980). *Phys. Rep.* **64** 172.

Blocki, J., Boneh, Y., Nix, J., et al. (1978). *Ann. Phys. (N.Y.)* **113** 330.

Bohm, D. and Pines, D. (1953). *Phys. Rev.* **92** 609.

206   References

Bohr, A. and Mottelson, B. (1975). *Nuclear Structure,* Vol. II, (Benjamin, New York).

Bohr, A. and Mottelson, B. (1981). *Phys. Lett.* **100B** 10.

Bohr, N. and Wheeler, J. (1939). *Phys. Rev.* **56** 426.

Bortignon, P.F., Broglia, R.A., Bertsch, G. and Pacheco, J. (1986) *Nucl. Phys.* **A460** 149.

Brandenburg, S., Borghols, W., Drentje, A., et al. (1987). *Nucl. Phys.* **A466** 29.

Bracco, A., Gaardehøje, J., Bruce, A., et al. (1989). *Phys. Rev. Lett.* **62** 2080.

Brechignac, C., Cahuzac, P., Carlier, F. and Leyguier, J. (1989). *Chem. Phys. Lett.* **164** 433.

Broglia, R.A., Barranco, F., Bertsch, G.F. and Vigezzi, E. (1994). *Phys. Rev.* **C49** 552.

Brown, G. (1988). *Phys. Rep.* **163** 167.

Chakrabarty, D., Dehnhard, D., Franey, M., et al. (1987). *Phys. Rev.* **C35** 1886.

Cohen, S., Plasil, F. and Swiatecki, W. (1974). *Ann. Phys. (N.Y.)* **22** 406.

Colò, G., Bortignon, P., Nguyen, van Giai, Bracco, A. and Broglia, R.A. (1992). *Phys. Lett.* **B276** 279.

Connerade, J. and Pantelouris, M. (1984). *J. Phys.* **B17** L173.

Deady, M., Williamson, C., Zimmerman, P., et al. (1986). *Phys. Rev.* **C33** 1897.

De Blasio, F., Cassing, W., Tohyama, M., Bortignon, P.F. and Broglia, R.A. (1992). *Phys. Rev. Lett.* **68** 1663.

de Heer, W., Milani, P. and Chatelain, A. (1990). *Phys. Rev. Lett.* **65** 488.

Engelbrecht, C. and Engelbrecht, J. (1991). *Ann. Phys. (N.Y.)* **207** 1.

Esbensen, H. and Bertsch, G. (1984). *Ann. Phys. (N.Y.)* **157** 255.

Esbensen, H. and Bertsch. G. (1984a). *Phys. Rev. Lett.* **52** 2257.

Fetter, A. and Walecka, J. (1971). *Quantum theory of many-particle systems* (McGraw-Hill, New York).

Frois, B., Bellicard, J., Condon, J., et al. (1977). *Phys. Rev. Lett.* **38** 152.

Gaarde, C., Rapaport, J., Taddeucci, T., et al. (1981). *Nucl. Phys.* **A369** 258.

Gaardhøje, J., Ellegaard, C., Herskind, B., et al. (1986). *Phys. Rev. Lett.* **56** 1783.

Gavron, A., Beene, J., Cheynis, B., et al. (1981). *Phys. Rev. Lett.* **47** 1255.

Gavron, A., Gayev, A., Boissevien, J., et al. (1987). *Phys. Rev.* **C35** 579.

Genzal, L., Martin, T. and Kreibig, U. (1975). *Z. Phys.* **B21** 339.

Goldhaber, M. and Teller, E. (1948). *Phys. Rev.* **74** 1046.

Goodman, C., Goulding, C., Greenfield, M., et al. (1980). *Phys. Rev. Lett.* **44** 1755.

Hanke, W., Mekini, N. and Weiler, H. (1985). In *Electronic structure, dynamics, and quantum structural properties of condensed matter,* ed Devreese and van Camp (Plenum, New York) 113.

Hase, W. (1976). In *Dynamics of Molecular Collisions*, ed. Miller (Plenum, New York) 121.

Heine, V. (1985). In *Electronic structure, dynamics and quantum structural properties of condensed matter*, ed. Devreese and van Camp (Plenum, New York) 1.

Heisenberg, J. (1981). *Adv. Nucl. Phys.* **12** 61.

Hernandez, J.J., *et al.* (1990). *Phys. Lett.* **B239** 3.78.

Hertel, I., Steger, H., deVries, J., et al. (1992). *Phys. Rev. Lett.* **68** 784.

Hinde, D., Ogata, H., Tanaka, M., et al. (1988). *Phys. Rev.* **C37** 2923.

Hohenberg, P. and Kohn, W. (1964). *Phys. Rev.* **136B** 864.

Kittel, C. (1968). *Introduction to Solid State Physics* (Wiley, New York), 3d edition.

Knight, W., Clemenger, K., de Heer, W. and Saunders, W. (1985). *Phys. Rev.* **B31** 2539.

Kawabata, A. and Kubo, R. (1966). *J. Phys. Soc. Japan* **21** 1765.

Konstanz conference (1991). *Proceedings of the 5th international meeting of small particles and inorganic clusters, Konstanz, Germany. Zeit. f. Physik* **D20** (1991).

Kubo, R., Toda, M. and Hashitsume, N. (1978). *Statistical physics* (Springer, Heidelberg), Vol. II.

Lamb, H. (1882). *Proc. Math. Soc. (London)* **13** 50.

Landau, L. and Lifschitz, E.M. (1970). *Theory of elasticity* (Pergamon, Oxford).

Landau, L. and Lifshitz, E.M. (1974). *Quantum Mechanics* (Nauka, Moscow).

Lipparini, E. and Stringari, S. (1989). *Phys. Rev. Lett.* **63** 570.

Lundqvist, S. (1983). *Theory of the inhomogeneous electron gas*, ed. Lundqvist and March (Plenum, New York), p. 149.

Mahan, G.D. (1981). *Many-Particle Physics* (Plenum, New York) p. 278.

Merzbacher, E. (1961). *Quantum Mechanics* (Wiley, New York).

Mie, G. (1908). *Ann. Phys. (Leipzig)* **25** 377.

Migdal, A. (1944). *J. Physics U.S.S.R.* **8** 331.

Negele, J. (1982). *Rev. Mod. Phys.*, **54** 913.

Nguyen Dinh Dang (1989). *Nucl. Phys.* **A504** 143.

Nguyen van Giai, Bortignon, P.F., Zardi, F. and Broglia, R.A. (1987). *Phys. Lett.* **B199** 155.

Pacheco, J., Broglia, R.A. and Mattelson, B.R. (1991). *Z. Phys.* **D21** 95.

Pines, D. (1963). *Elementary excitations in solids* (Benjamin, New York) 209.

Pines, D., and Nozieres, P. (1966). *Theory of Quantum Liquids,* (Benjamin, New York).

Raman, S. Malarky, W., Milner, C., et al. (1987). *At. Data and Nucl. Data Tables* **36** 1.

Rapaport, J., Kulkarni, V. and Finlay, R. (1979). *Nucl. Phys.* **A330** 15.

Ring, P. and Schuck, P. (1980). *The nuclear many-body problem* (Springer, Heidelberg).

Sasao, M. and Torizuka, Y. (1977). *Phys. Rev.* **C15** 217.

Satchler, G. (1971). *Nucl. Phys.* **A195** 1.

Scheidemann, A., Toennies, J. and Northby, J. (1990). *Phys. Rev. Lett.* **64** 1899.

Schroeder, W. and Huizenga, J. (1984). *Treatise on Heavy Ion Science*, ed. Bromley, Vol. II (Plenum, New York), 115.

Selby, K., Vollmer, M., Masui, J., et al. (1989). *Phys. Rev.* **B40** 5417.

Serra, Ll., Broglia, R.A., Barranco, M. and Navarro, J. (1993). *Phys. Rev.* **A47** 1601.

Sharma, M., Borghols, W., Brandenburg, S., et al. (1988). *Phys. Rev.* **C38** 2562.

Shlomo, S. and Bertsch, G. (1975). *Nucl. Phys.* **A243** 507.

Speth, J. and van der Woude, A. (1981). *Rep. Prog. Phys.* **44** 719.

Suzuki, T. (1973). *Nucl. Phys.* **A217** 182.

Steinwedel, H. and Jensen, J. (1950). *Z. Naturforsch.* **A5** 413.

Tassie, L. (1956). *Aust. J. Phys.* **9** 407.

Thoennessen, M., et al. (1987). *Phys. Rev. Lett.* **59** 2860.

Ullah, N. and Rowe, D. (1970). *Nucl. Phys.* **A163** 257.

Vautherin, D. and Brink, D. (1972). *Phys. Rev.* **C5** 626.

Wambach, J. (1988). *Rep. Prog. Phys.* **51** 989.

Wang, C., Pollack, S. and Kappes, M. (1990). *Chem. Phys. Lett.* **166** 26.

Weyhreter, M., Barzick, B., Mann, A. and Linder, F. (1988). *Z. Phys.* **D7** 333.

Youngblood, D., Rozsa, C., Moss, J., et al. (1977). *Phys. Rev. Lett.* **39** 1188.

Yannouleas, C. and Broglia, R.A. (1991). *Phys. Rev.* **A44** 5793.

Yannouleas, C. and Broglia, R. (1992). *Ann. Phys. (N.Y.)* **217** 105.

Yannouleas, C., Vigezzi, E. and Broglia, R. (1993) *Phys. Rev.* **B47** 9849.

Index